THE CHOICE

PETER O. CHILDS

ARPress
ILLUMINATING IDEAS,
EMPOWERING VOICES

ARPress
45 Dan Road Suite 5
Canton, MA 02021

Hotline: 1(888) 821-0229
Fax: 1(508) 545-7580

Ordering Information:
Quantity sales. Special discounts are available on quantity purchases by corporations,associations, and others. For details, contact the publisher at the address above.

Printed in the United States of America.

ISBN-13:   Softcover   979-8-89356-909-4
                eBook   979-8-89356-910-0

Library of Congress Control Number: 2024901288

TABLE OF CONTENTS

This book is dedicated to you.

"There is no religion higher than Truth"
-(Theosophists)

*"We are stardust; we are golden, and we've got
to get ourselves back to The Garden."*
- (Joni Mitchell)

PROLOGUE

The following is an attempt to put forward what I have been taught and do believe is really happening at this uniquely threatening but much more than equally promising juncture in human history. It is huge; vastly larger and more important than almost anything we're accustomed to think about. Nonetheless, it's happening and we desperately need to understand what it is and what to do about it.

This discussion is primarily aimed at those who like to think; who have discovered what an amazing tool their mind is and how it enables us to "figure things out". But I want to emphasize that one does not have to intellectualize these matters in order to understand them. Many of us intuitively understand the most important issues in Life; they live good lives simply because that is what feels right to them.

So the approach taken here is absolutely not the only avenue to truth. But it's one of the best; on that basis let us proceed.

PREFACE

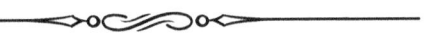

In writing this dissertation, "The Choice", I face a choice at the outset and it's a difficult one. The whole intent of this effort is to make clear that regardless of the chaotic conditions that we now face on Planet Earth, the real journey that we are on is certain to have an outcome literally infinitely more wonderful than anything that, owing to our normal habits, most of us can yet even allow ourselves to dream of. Those normal habits must now change and change radically but in order for that to happen we must see clearly what is actually going on around us and within us, and that will be the subject of this discussion.

The difficulty here lies in the fact that in order to understand the central, fundamental choice that we must each now make we have to understand clearly what that choice is; what we are choosing between, and that is nothing less than Good and Evil or in other words, Right and Wrong or according to God's will or against God's will (we'll dive into what that is; what the word "God" means, a little later).

In order to obtain that clarity, it is essential that we see the Evil, so long as it exists, as well as the Good in order that we at last understand what they both really are. In the exponentially accelerating rush of human affairs, most spectacularly in the last and present centuries, all forms of Right and Wrong are intensifying; coming right up in our face in order that we may see clearly the vital difference between them and now, at last, choose once and for all which master we will serve. We have been attempting to serve two masters throughout our history, ever since the point where we first accepted Wrong in our lives, which is essential history that we will discuss in depth.

So, much as I would prefer to set forth only the good and the beautiful (which I must incessantly emphasize is the whole point of this treatise) I feel obliged to deal with the bad and the ugly, which at this point in our history I doubt that I have to point out is very much in the ascendent. So let's get that out of the way so that we can put it in its proper place. Let us awaken to how much larger our life is than we have taken it to be; let us expand our awareness of the greater realities that have always been around us and within us waiting for us to open to them, and let us realize the indescribably wonderful promise that they hold for us and for all of God's Creation.

WHAT HAS GONE WRONG

WHERE TO BEGIN?! Let's start with the single most encompassing problem that we face; the one that trumps all the other enormously significant challenges that are now crashing down on our heads: climate change. If we do not solve the problem of climate change none of the other problems will matter, because in all likelihood we will not be around in any condition to deal with them. Why? Because (and we must be perfectly clear about this; we must not continue to turn our heads away in ignorance and denial) the complex of problems that center on climate change are so severe that they now actually threaten life as we know it on this planet.

In 2018 the United Nations Intergovernmental Panel on Climate Change announced that we might have as little as twelve years left to avoid "catastrophic" global changes as our planet warms, its great polar cold-storage centers diminish, and the resulting altered weather patterns wreak havoc of various kinds such as increasing hurricanes, floods, fires, and droughts, one of the most severe impacts of which will be in our ability to grow food. A recent study indicates that forty percent of the human race live in areas that will become uninhabitable if we go beyond the 1.5-degree temperature increase limit set at the Paris Accords, a limit that we're rapidly moving to exceed. We have a new term gaining currency: "climate refugees"; already tens of millions of us (they are us) have fled these impacts every year since 2008 and that number will only increase unless and until we deal with this tremendous problem, which should be front and center every day in every news medium but which is still largely ignored in our mainstream media, our politics, and our everyday conversation. Instead of seeing the reality here and dealing with it we turn our heads away from this enormous and immediate threat. Why? This is something that we need to understand and correct; it is worth examining closely.

The wonderful Greta Thunberg was asked recently to comment on the idea that most people realize the enormity and immediacy of the danger into which we've allowed ourselves to drift but that it's just too big and too scary to face so we tend to turn our attention away from, instead of toward it. She replied that she doesn't agree with that, that in her highly educational travels she's come to the conclusion that too few people have any significant awareness of what's going on around them. The reason for this is that 1) We're too busy; our attention is commanded by our immediate needs, so many of which are currently so hard for so many of us to meet. 2) Who tells us the truth about such things? Not our parents, not our friends, not our schools, not our churches, not the media that most of us listen to; indeed, many of the sources that so many of us (roughly half!) rely on for information actually give us a steady diet of terribly dangerous falsehoods that we accept not because we've checked their veracity but because they satisfy our emotional needs. 3) Tremendously powerful entities deliberately and very effectively shape our view of things. From our shamefully distorted political parties, who have given over their fundamental responsibility to the citizenry's welfare instead to corporate lobbyists and political donors, to the media who have done the same thing; making money their bottom line rather than dissemination of truth, to our educational institutions who above all should be teaching critical thinking but who fail miserably to do so, enormously powerful forces shape what we see, think, and feel.

We need to awaken to how seriously our media have failed us. Their main functions are to educate us, entertain us, and hopefully to inspire us (which is the highest form of education). In a democratic society it is obviously vital that the voting citizenry be truthfully and adequately informed regarding all matters that significantly affect them, in order that they be able to cast a responsible vote. But our media are failing badly in this essential activity. Our most important newspapers and our TV newscasts seem to be only marginally more concerned with any deep analysis of reality than our TV "reality" shows, which bear little more resemblance to actual reality than green cheese on the moon. Then we have, in great force, "right wing" media who deliberately fill the air with outright absurdities and lies of the most dangerous kind, pouring them into minds that are unaccustomed to critical thinking, to such harmful effect that for one excellent example we in the United States of America, the "shining city on a hill", actually allowed such a wretched specimen of humanity as Donald Trump to become President, and still support him in horrifying numbers.

To the brothers and sisters who have been taken in by this man and his way of seeing things I intend no offense. But I must insist that this is the truth. I say what I say about Donald Trump not because I see him as less than a fellow child of God (The placard I would have waved at demonstrations if I lived anywhere near where they happen would have said "God Bless Donald Trump! Bring him wisdom! Bring him love! Please hurry!!!") but because I have studied him carefully. Obviously he is an important figure but unfortunately not by any stretch of the imagination as a wise, loving, knowledgeable leader but as a stunning example of how easily so many of us can be hypnotized into accepting terribly wrong things, and how willing some people in search of power are to take advantage of this regrettably common weakness on the part of the general public who are so unused to critical thinking, i.e. the proper use of God's second greatest gift, our mind (third is our body, first is our heart).

A recent study of the January 6 insurrection shows that the primary motive for the action (whose participants consisted of only ten percent who were affiliated with "right wing" extremist groups and roughly forty percent of whom were business owners or white-collar employees), was concern (fear) over "replacement", i.e. the realization that non-white human beings are increasing their percentage of our population, which is seen as problematic because of the view of the United States of America as a white nation, started by Europeans (nobody need think of the Indians), and the feeling is that it should remain a white society, uncontaminated by what is felt to be lesser members of the human race. More than half of the insurrectionists who have been arrested do not come, as one might predict, from predominantly "red" counties but from blue ones where Biden won; where former minorities are increasing their numbers and are perceived by many people to be competing with the established white business community.

If we stand back even a little bit the problems with this attitude are self-evident but sadly and with terrible consequences, Donald Trump and his followers refuse to stand back even an inch. Thus they are unable to see what they are doing ("Forgive them, for they know not what they do") how seriously harmful are the things they are enabling both in government, where they maintain a greatly outsized presence, and in the actions of emotionally driven groups like the Proud Boys, Three Percenters, et al.

Certainly, we should do our best to love the doer while we hate the deed, but we must not ignore these deeds; they are so harmful. You don't blame the child for playing with matches but you don't just let them do it, and you certainly don't give matches to the child to play with (as in allowing Donald Trump to assume the Presidency). So if it is claimed that Trump actually did win the 2020 election; that it was fraudulently stolen from him, let this be proven in the light of day. If the "Commie Democrats" are actually worshiping a Satanic cult that uses children's blood for vile purposes let this be proven, not just wildly claimed. This will not be possible because it has already been proven in the full light of day that Biden won the election, and there exists not a shred of evidence to support the QAnon lunacy. These madcap theories that are being tossed about and so easily accepted can only exist in the dark; they cannot stand the light of Truth.

But those of us who go along with these enormous and so harmful falsehoods are not looking for the truth; they're attempting to satisfy their emotional needs, so many of which (e.g. needing to feel superior to other people) are, to put it in the mildest terms, harmful rather than helpful. This must change. We need desperately now to see the truth of things, and in order to do that we must be open and honest with each other. If we can do this we will be very glad that we did; we are all children of God, members of one indescribably wonderful family and much as it might seem otherwise at this tempestuous time, we are all bound for glory, as I hope to make clear.

In order to find any in-depth presentation of important matters one generally has to turn to alternative media whether in print, on radio or (surgically) in social media and far from being recognized as the best sources of information that they are, the alternative media are generally thought of as too "far out" to pay attention to. So we remain largely and very dangerously ignorant.

This is a shameful state of affairs. Rather than our citizenry being alert, inquisitive, and demanding of truth in what is being said to us, we have allowed our minds to be filled with all manner of falsehood, which is reflected in our votes (and in our not voting), with the result being the political chaos that we now enjoy, not to mention its widespread effects on our nation and our planet through the stunning incompetence of so many of our elected representatives.

How can it be that we allow ourselves to be affected this way? We are supposed to be in control, through our votes, over our government, our economic system, and in general over every aspect of our society, are we not? Plato: "I need no longer hesitate to say that we must make our guardians philosophers. The necessary combination of qualities is extremely rare. Our test must be thorough, for the soul must be trained up by the pursuit of all kinds of knowledge to the capacity for the pursuit of *the highest—higher than justice and wisdom—the idea of the good*" (emphasis added for reasons which will become apparent).

Very likely our second most dangerous problem is nuclear proliferation. This is another of those hugely significant realities that we generally just ignore, which only allows the problem to become worse. The Bulletin of Atomic Scientists has been keeping a "doomsday clock" since 1947 that registers their assessment of the current level of danger from nuclear fission, whether from accidents, radioactive waste, or outright nuclear war; the closeness of the hour to midnight indicates the threat level. In 2020 they set it at one minute and forty seconds to midnight, the highest level they've ever recorded. Why? For several reasons, but what they boil down to is things like the fact that we should never have developed nuclear fission without the ability to deal effectively with radioactive waste, which ranges from low-level radiation from sources such as hospital and laboratory instruments to the thunderous radiation from the accidents at Three Mile Island, Chernobyl, and the Fukushima plant in Japan, or that which has resulted from our bomb tests (two of which wreaked unimaginable horror on the Japanese people when we tested them on humans). The Fukushima disaster is by no means being effectively dealt with; enormous amounts of water have been used to cool the wreckage, which includes three melted cores furiously festering underground with as yet no way to deal with them. The irradiated water has been stored in massive tanks which are reaching capacity, so the plan is to release the treated but still contaminated water into the ocean, which is madness.

The radiation from nuclear waste can last tens of thousands of years; we know this perfectly well but we still have no way to effectively store or dispose of that waste. Check for just one example, the underground plume of radioactive water that has been steadily moving toward the Columbia River from the decommissioned Hanford nuclear site in Washington State. And keep an eye open for your government trying to convince you to stash radioactive waste in your backyard.

Then, with regard to nuclear weapons, there have been a frightening number of accidents (nukes lying in a wrecked submarine at the bottom of the ocean, bombs falling out of bombers, accidents in missile silos, etc.) some of which came very close to potential disaster. Not to mention how close we have come to a "nuclear exchange" with other nations; there have been a number of close calls due to misunderstandings of one kind or another and during the "Cuban missile crisis" a Soviet submariner actually refused an order to fire a nuke. How many of these bullets can we count on dodging, especially as we continue to expand rather than reduce our nuclear arsenals while at the same time ratcheting up international provocations that increase the danger of nuclear war?

Aside from the threat of nuclear holocaust we must face the enormously significant reality of how far the "military/industrial/ congressional complex" (Ike's original term) has developed along the very lines against which he warned us. Far from having seen the danger, we have turned our heads away and allowed the MICC to have its way (we don't even audit the Pentagon!), which has led us into what is now commonly called "never- ending war" that has snatched away unbelievable amounts of our tax money that desperately needs to be spent elsewhere on real human needs, and which has increased our fear of our "enemy" brothers and sisters around the world, upon which fear the MICC depends. It's so much easier to demonize an "enemy" than to understand them, particularly when you are making fortunes from so doing. Never was there a clearer case of "follow the money"; can you spell "lobbyist"? The MICC has an army of lobbyists constantly ensuring that Congress does what they want, strongly assisted by donations to members of Congress who will support them, and by the whole system of PACs, think tanks, etc. with which they maintain their grip on our government and to a regrettable degree, our minds. Of course Russia is our enemy. Of course China is our enemy. Of course Venezuela is our enemy. Of course Iran is our enemy. We have to defend ourselves against the constant huge threat that they pose to us. Really?

In our entire two hundred and forty-four-year history we have enjoyed only sixteen years without a war. We need to thoroughly assess our normal attitude toward our military services, the functions that they actually perform, and the real need or lack thereof for those functions. Our default assumption is that whatever our military does is necessary for our defense, not because it really is necessary but because that idea is drummed into our heads by the Military/ Industrial/Congressional Complex and the media that almost never question it. This has been the case for time out of mind, like so much else that we believe out of pure habit, not at all from having seriously examined it. Let's examine it.

Without question, it is heroic for a person to be willing to risk life and limb to protect their loved ones, be they family, friends, or their nation from whatever significantly endangers it ("all enemies foreign and domestic", let us not forget; think "whistleblower"). We take it for granted that whenever our government sends out troops it does so in response to a real threat; a threat that can only be dealt with through military force. And in this wicked world such has indeed been the case throughout human (and animal) history; evildoers have posed real and substantial threats to the wellbeing of their fellow humans, who have had to respond by defending themselves, with the soldiers in the field willing to face the horrors of war. Their heroism is real, and deserving of every honor that we accord it.

OK. How about the enemy? He too believes that he is fighting for a righteous cause; he sees you as the enemy, the bad guy doing wrong and threatening him, which you certainly are. On what basis are we to demean his heroism while we recognize and praise our own? Ulysses S. Grant recognized the importance of this question when, after accepting General Lee's surrender, he said "I felt like anything rather than rejoicing at the downfall of a foe who had fought so long and valiantly, and had suffered so much for a cause, though that cause was, I believe, one of the worst for which a people ever fought, and one for which there was the least excuse." What is God supposed to feel, watching his children tear each other to pieces, each of them claiming all the while that God supports their cause?

The truth of the matter is that a soldier fighting with pure motives coupled with a clear understanding of the cause for which he is fighting is the exception, not the rule. The first rule of a soldier is to obey without question the commands of his superiors. This is an obvious necessity; democracy in military operations is impossible. But that makes it not just possible but highly likely that soldiers will be fighting not because of a heroic understanding of what they're protecting but simply because they happened to be in the Army when a conflict arose (far too often because they needed a job, not from high and noble motives) or because they were drafted, or because they accepted a call to war without seriously questioning it. Most combat veterans that I know agree that they were fighting not for particularly noble principles but to stay alive and to keep their buddies alive.

It can be persuasively argued that World War II was the last war the United States fought that was morally justifiable; that actually was a necessary defense against a massive evil that left no alternative to war. But the man who led the winning forces in that war and who subsequently became President, left us with a searingly perceptive warning about how easily we could be led into war for anything but noble reasons. He warned us against the "military/industrial complex" that was based not on realistic assessment of threat, not on high moral purpose, but on the enormous profits to be made from the manufacturing and sale of instruments of death. "Every gun that is made, every warship launched, every rocket fired signifies, in the final sense, a theft from those who hunger and are not fed, those who are cold and are not clothed." "Is there no other way the world may live?" "This is not a way of life at all, in any true sense...it is humanity hanging from a cross of iron." "We must guard against the acquisition of unwarranted influence, whether sought or unsought, by the military-industrial complex."

We failed to heed that warning. The MICC has steadily increased its grip on our minds and our purse to the point where even now, having failed in virtually if not literally every military endeavor we pursued since WWII, we still give the Pentagon more than half of our discretionary spending while real and increasingly desperate domestic needs go unmet. We unquestioningly allow our government to establish nearly eight hundred military bases in seventy countries, and when our government announces a new military operation (should they choose to even announce it) far too few of us question it; we just go along with it.

Tragically far from our minds are the words of the Sixth Commandment, which said "Thou shalt not kill." Period. Not "...except when your arms manufacturers and dealers can lobby Congress to sell more weapons to anybody who will pay for them" or "...except when your government decides that it needs to shore up its control of Middle Eastern oil, or Latin American resources, or African resources," or anything else. "Thou shalt not kill." Period.

When, if you stop and put your mind and heart to it, was it ever right to shoot pieces of lead into each other's bodies? When was it ever right to stab each other, bludgeon or bomb each other, or otherwise ruin each other's lives? When, really, was it ever right to do anything to cause each other pain of any kind?

Why do we almost universally accept this state of affairs? Why can't we see where it must eventually lead if not radically changed? It doesn't take an Einstein to see that if I consider you to be my enemy I need to be able to defend myself against you, so I feel obliged to have a stronger military capability than you do or at least one strong enough to deter you. But when I develop any significant advance in that capability, you feel obliged to deal with the situation by developing new weapons of your own (not because you really need to but because it gives you an excuse to make more money and flex your muscles). This constant leapfrogging guarantees two things; that more powerful weapons (of whatever kind come to hand) will steadily appear, and that the dangers of using them accidentally or deliberately will constantly increase. True, nobody is going to try to leapfrog over the United States' and Russia's nukes (or China's, should they decide to escalate to that degree); nobody could afford to because we have so many bombs. In the U.S. we spend more than half of our discretionary spending on our military; we're currently in the process of "modernizing" our nukes to the tune of between one and a half and two trillion dollars (yes, trillion). Nobody else can match that level of insanity but for any nation striving to be a major player in world affairs or even just to protect its own interests it makes perfect sense to develop a nuclear military capacity; to join the "nuclear club" and thus gain a seat at the table or at the very least to be left alone. So on we go, steadily courting catastrophe.

This is not a future to which reasonable people can look forward with pleasure. Apart from the massive amounts of our tax money that we dole out to the MICC, in our reckless insanity we are now (thanks to Donald Trump) weaponizing space with a whole new branch of our military, the Space Force.

Just for the fun of it, suppose that we had a Department of Peace. Suppose that they operated on the basis of the realization that each and every one of us almost certainly does some things right and some things wrong. The things that we do wrong threaten each other, so we have Departments of Defense (ours used to be more forthrightly called the War Department), which translate our concern (fear) into military action. Understandably, but the function of the Department of Peace would be to look not for reasons to turn away from each other but to turn toward each other in order to build bridges of understanding and sympathy rather than chasms of ignorance, mistrust, and hate. For example, what if we were to set aside for the moment the serious failings of Russia's government (and our own) and take a more open attitude toward Russia? What if we were to realize that for centuries they have been one of the most important nations on earth? That the people who fill their streets are virtually indistinguishable from the people who fill ours? That their nation has disintegrated twice in living memory, and that they'd prefer that that not happen again; that they would like to continue to be an important member of the community of nations? That their fear of us is easily understandable when we see the truth about how we have treated them; this is not the place to get into that in detail but consider for example how Gorbachev agreed not to oppose the reunification of Germany in exchange for George H.W. Bush's promise that NATO would not advance "one inch closer" toward Russia's borders (their "sphere of influence"), which promise we then treated exactly like our treaties with the American Indians, every one of which we broke. NATO now brings massive military exercises right up against Russia's borders; we didn't care for that when they did it to us in Cuba, but if we do it to them it's OK? Russia shouldn't object? These are the kinds of honest and necessary questions that a Department of Peace would ask and act upon, and the world would be very much better for it.

We cannot afford to continue wasting half of our dollars on the machinery of death and destruction. We do not need to; that hideous falsehood is ginned up by the MICC, spread by our politicians and our media (even our churches and schools), and taken for granted by most of us, who do not spend much time examining such realities. This must stop, or we will dearly wish it had.

Also high on the list of enormously serious problems we face is the astonishing division that we have allowed to develop within our society. In every society there is necessarily a basic social compact; a set of rules, spoken or unspoken, that recognizes basic standards of human decency in accordance with which we must treat each other if we expect our society to function successfully. That social compact has been badly broken. Rather than recognizing our disagreements as normal human differences in perception and dealing with them like adults, listening openly to each other, learning from each other, defining our common interest, compromising and working together to find the best ways to meet that common interest, we now shut each other out, make no attempt to understand each other, hurl insults at each other and even face each other with guns, evidently forgetting that we did that once before with ghastly results in the Civil War.

This terrible damage to our society has at every turn not been recognized, faced, and dealt with properly; it has been strongly encouraged by those who should have fixed it; our elected representatives. The President of the United States (Donald Trump) has consistently encouraged this kind of behavior; our Congress, led by the Republican Senate majority but all too often with the meek acceptance of most of the rest of Congress, has failed miserably to remedy this cancer. One need not be a statistician to understand this; it's perfectly obvious to anyone who follows the news that Donald Trump's exhortations led directly to violence (and continue to do so, with Asians the latest victims because of Trump's nonsensical self- exculpatory "China Virus" claims). To say nothing about the stunning numbers of deaths that his profoundly criminal negligence with regard to COVID 19 has caused. And not to mention his ginning up the Jan. 6, '21 insurrection in which his followers actually attempted to overthrow the legitimate government of the United States by force.

This division amongst us is deliberately and very effectively used against us; to herd us into groups whose energies are squandered on fighting each other rather than communicating effectively with each other, which would threaten the supposed interests of those who benefit from stoking these fires. They are terrified of our actually communicating with each other, listening to each other, realizing our common interest (we share so much more than what divides us!), and acting on that interest. It is highly regrettable that the powers that be fail to realize that this common interest is theirs too; that for example, if we fail to deal with climate change we'll all be in the same sinking ship.

There is a hugely significant aspect of this division that must be seen and understood. As observed above, we humans constantly see things from different points of view and thus disagree about what it is we're looking at. This is normal; two people (or groups) can see something very different and both can be correct; as with the blind men and the elephant, one may have the animal by the tail and reasonably conclude that the thing is like a broom. Another blind man may have hold of a leg, and understandably feel that the thing is quite like a tree. Both of them, if they're willing to listen to each other, can learn from each other, expanding their understanding of what an elephant actually is. But all too often, especially in times like these when we're at each other's throats, we don't listen; we simply demonize each other like children squabbling in a sandbox. We're unaware of our blindness; we think that we know what the elephant is when we're actually only aware of a part of it. Then when we turn away from each other rather than listening to each other we deprive ourselves of whatever we could have learned from each other; classic self-defeating behavior.

But very importantly, we must recognize that our current situation does not consist of two groups of people looking honestly at the reality around them from different angles; it is very much and increasingly a case where one group is attempting to see things as they are, and the other is insisting on seeing things as they are not. And not just seeing things wrongly; vigorously insisting (without evidence; in fact, in direct conflict with the massive preponderance of evidence, as with Trump's astonishing ongoing insistence that he, not Biden, won the election) that the other side is actually wrong;

that they are the ones seeing things as they are not; who are causing all our problems and who must be dealt with (severely of course [Trump-" Just shoot them!"] to ensure that they stop).

The language of truth is being used in pure Orwellian fashion to utter falsehoods; very important falsehoods, e.g. that yes, there is a terrible division in our society but that that division is entirely the fault of the very people who are trying to make things right, such as the protesters against the systemic racism that so many of us are beginning to realize has permeated our society from its very outset. That our political divisions are between the righteous Republicans and wholly wrong "communist socialists" (Progressives and to a degree, Democrats) who oppose them. That COVID 19 is not a serious problem requiring unusual measures such as mask-wearing, social distancing, and mass vaccination; that that idea is a hoax perpetrated by the "radical left", the Chinese, or whoever happens to be your enemy du jour. So far have these people departed from reality that utterly mad ideas such as QAnon's insistence that a Democratic cabal of Satan-worshipping pedophiles is running a global child sex-trafficking ring and plotting against our government are actually gaining currency (a recent poll showed that fifteen to twenty percent of us [fifty-six percent of Republicans] agree with this lunacy!). We are unwittingly demonstrating something very important: that once we accept anything wrong we're in mortal danger of accepting anything wrong.

No better example exists to demonstrate how terribly harmful such illusions can become when we translate them into group action, than the current state of the Republican Party, which has sold its soul entirely; they have become, as the exemplary Noam Chomsky has been saying for years, "the most dangerous organization in human history". They have, through thoroughly anti-democratic means (e.g. the Electoral College, gerrymandering, anti-voting laws, etc., not to mention the Satanic Fox News and the money of the likes of the Koch brothers with their spigot opened wide by the Supreme Court's "Citizens United" travesty), acquired power vastly out of proportion to their numbers. They have done this deliberately ever since Reagan (Lee Atwater, Karl Rove, Newt Gingrich) kicked the process into gear; they are well aware of how successful they have been, and they are doing everything they can to snap the slender thread by which the power of the Democrats presently hangs in 2021. They have no more interest in the wellbeing of their fellow Americans than they do in the wellbeing of the environment or, for that matter, of our brothers and sisters in other lands. They are interested in the power, money, and prestige that are heaped on them by their hypnotized cult followers. They are deliberately and recklessly destroying democracy in the U.S.A; they know full well that as Trump said right out in the open (as usual) that if everybody got to vote "you'd never have a Republican elected in this country again". They are extremely dangerous and anything but toothless. If the Biden administration can manage to get enough programs through this Congress (they must wake up and play hardball; do away with the filibuster, etc. so that the Republicans can't prevent their programs from being carried out) then the average American might conceivably wake up to the fact that government can and should be looking out for the interests of the many, not the few, and we might be able to rescue the American Experiment. It's completely understandable that so many of us are disillusioned with our government, which has not only been ignoring our interests but working directly against them for decades, but it's shocking and worrisome that we're so ignorant as to the causes of our government's behavior. It's time for the truth to come out. It is time for the American people to realize what has been done to them, by whom, how and why, and to put a stop to it.

WHY DO WE BEHAVE THIS WAY?

THE CONTENTION HERE is that what is going on is normal human dynamics greatly exaggerated by circumstances. Human beings are far more motivated by emotional needs than by rational thinking or careful observation of facts. We have a fundamental and wholly appropriate need for certainty; for a sense of things that won't fail us, that won't allow the rug to be pulled out from under our basic view of Reality. But so often events force us to realize that our sense of things was simply wrong, and that's always a shock because it threatens our very sense of Reality, and short of death (which we will unmask later) there is no greater threat than that.

We do not need to defend our views when they are actually correct; in accord with the real world. You don't have to insist that two and two equal four; they just do (God arranged things that way). You can rely on it; you can feel secure in your sense of it, and here is one of the most important principles that we need to learn: it is entirely appropriate that we yearn for certainty (the only alternative is chaos) but the only real certainty that exists or could ever exist is a living awareness of Reality. Anything other than Reality is illusory and thus uncertain; subject at any time to being snatched away. Truth (Reality) has been referred to as "that which does not go away when you stop believing in it". But why would we stop believing in Truth? And why would we settle for a matter of belief rather than of knowledge, i.e. the real certainty that we so rightly and deeply crave?

What we're talking about here is Truth vs. Falsehood, which will be the subject of our entire subsequent discussion. Suffice it to say here that we humans have the ability to create reality ("made in the image of God") but so much of the time we misuse that ability and create wrongly, which is the source of all our problems; in fact, it's the only problem we've got. Every problem we've ever had is directly traceable to our misuse of our creative power; we'll examine that more deeply later.

So onward we hurtle toward the brink of a unique existential Cliff; the global "catastrophe" that the U.N. warned us would be caused by climate change alone, to say nothing of the other historically unique threats that we have brought upon ourselves. We are not in this mess because of what we have done right; it is because of what we've done wrong, the consequences of which have built over time as we continued to make the same old mistakes that we pretend are unavoidable ("human nature") with ever-increasing power and which now threaten to overwhelm us. The situation certainly is dire but it is not hopeless; if we can stand back and look at it objectively we can see that all its complexities are subject to certain fundamental principles that are not complex at all, and which if and when we understand them can not only save us from The Cliff but launch us into a whole new octave of Life more wonderful than anything we've allowed ourselves even to dream of.

IGNORANCE

THE MYSTICS OF this world have long insisted that "the only sin is ignorance". There is no condemnation in this statement; sin is defined as "missing the mark", as in shooting at a target. In this sense it's a purely mechanical statement; you hit the bullseye, you get the prize; you miss it and you don't. Knowledge is power; if you see things as they really are you can operate successfully in life. If you see them other than as they are you will have problems. And that in a nutshell describes the human condition; we constantly try to operate on the basis of perceptions and assumptions that simply are not true; that do not comport with reality. Then we wonder why things do not turn out well.

Again, there is no condemnation in pointing out that we fail so spectacularly; the results of that failure are condemnation enough in themselves. We fail for certain reasons, all of which boil down to ignorance, but we do not understand this. We just continue to fail in the same time- honored ways, telling ourselves "that's life" by which we mean that it's inevitable, which it is not and never has been, but so long as we look at it that way we make it functionally impossible to change because we won't even imagine that it could be otherwise and ask how to bring that about.

This is still a nation and a world in which, if we will, we can do what we're here to do. We can break free from "the way it's always been done". We can focus our energies where they should be focused; where they always should have been focused. For example, freedom of the individual has always been the core of the appeal of the United States of America, but we have had far too narrow an interpretation of the concept, generally taking it to mean the freedom to do as we pleased regardless of whether it was right or wrong, or of the consequences either to ourselves or to other creatures. If we realize what it really means to be free; that it means understanding that freedom and responsibility must not be separated and that it is our responsibility to live only with Good, then we have everything we need to move as fast as humanly possible in the right direction.

The fundamental character of evolution is the increase in intelligence displayed by individual life forms. A plant is more intelligent than a rock; an animal is more intelligent than a plant, and a human is more intelligent (as far as we know) than any other life form on this planet. We have demonstrated a wonderfully high degree of intelligence, from our material inventions to our scientific discoveries, to our philosophical realizations. How then can it be that with all this intelligence we have brought ourselves to this critical state of affairs? We have made it perfectly clear that we have the ability to save ourselves, to save the world; to understand what we are doing and why we are doing it. So why aren't we doing that? No rock, plant, or other animal is going to save the world; it's up to us.

We aren't doing it not because we lack the ability but because we lack the will. The reason why we lack the will is because of our ignorance, and one of the most important things of which we are woefully ignorant is faith. We don't understand what faith is; how essential an ingredient it is in right living. It all goes back to The Fall (we'll get into this); ever since then we've been living "by the sweat of our brow", in constant fear of not having enough (you never have enough because the more you have, the more you have to lose) and preoccupied with our ceaseless efforts to get what we need and then, hopefully, what we want.

Life is what it is, whether or not we see it rightly, and when we do look at it rightly we increasingly realize that it is, at the root of all its complexity, perfectly simple and wonderful beyond the power of words to tell. In fact, it is our destiny to awaken from our nightmarish confusion; to be released from this prison of wrongness and pain and to enter into real Life, which is not just better; it is perfect and eternally so, wholly composed (when we get free) of an infinitely expanding series of ongoing good things. Let's consider the matter in some depth.

WHAT WILL GO RIGHT

"THE ONLY SIN is ignorance." Ignorance of what? Of the answers to the most important questions we can ask, of which there are four: "What am I?" "Where do I come from?" "Where am I going?" "What's going on?" Almost none of us ever bother even to ask these questions, let alone to give them any serious attention, for a very good reason; we do not believe that it's possible for us to answer them and that attempting to do so would be a waste of life's precious time. A football game has far greater attraction for us than "unscrewing the inscrutable" so we don't try, with the result that we bash on in ignorance, constantly paying the price of that ignorance without realizing it, and with that price steadily increasing until it now threatens to overwhelm us, which was inevitable because if we continue to do wrong things with ever increasing power the results must, sooner or later, build to the point where they crush us. Which in a nutshell is precisely what is happening now; we are not in this mess because of what we've done right but because of what we've done so wrong for so long, and the situation has become so serious that unless we wake up, see what's really happening, and take appropriate action we'll knock the human journey into a serious backward eddy (nothing can stop that journey, as I hope to make clear). So let's wake up. Let us, for a tremendous change, seriously consider those Four Great Questions, using our God-given eyes, minds, and hearts.

Let's start with our history; not the extremely limited account that we humans have produced since we learned to write it down (and largely infused with opinion rather than fact), but the vastly extended version that is put before us in the Bible and which is true and important at a depth that we seldom realize unless we are familiar with the teaching of our mystics. This history is, we must realize, thickly covered up by layers of varied understanding on the part of those who wrote, translated, and interpreted the Bible but it is possible to apply x-ray vision, as it were, to look through these layers and get to the solid gold underneath. The Bible is the extended history of the human journey; our origin, the Great Mistake that we made, the consequences of that Mistake, and our eventual emergence ("salvation") from that regrettable state of affairs. Here is what it boils down to:

Before there was anything there was not nothing; there can be no such thing as nothing. There was That Out of Which Every Thing Came, which far from being nothing was and is undifferentiated everything. This great creative Reality we call God, or the Creator because everything came "out of" it; it created everything. It should go without saying that all our religions are worshiping the same Creator, whether we call it God, Allah, Jehovah, or whatever other term we care to use. It is self-evidently ridiculous that there could be one Creator for Christians, another for Muslims, a third for Jews, etc. This is something that we should wake up and agree on; it is so obvious and it would be such a beneficial thing in terms of drawing us together rather than driving us apart, which our clinging to our religions rather than understanding them so often does. I've contacted the Vatican twice, begging the Pope to convene a meeting with the leaders of the world's religions, recognize this vital fact, and together announce their agreement to the world. They have shown no interest.

At this point it is essential that we realize how something that is not perceptible by our five senses is real; how anything that isn't "thingy" can be real. Well, ideas are just as real as shoes, and they're not subject to perception by our physical senses; can we at least hypothesize that prior to things appearing in a material form they existed as ideas in the mind of That Which Created Everything, just as everything we create exists as an idea in our mind before we bring it into material existence? This is a very useful hypothesis because it gives us a comprehensible context for the history that the Bible gives us regarding what we are, where we came from, where we're going, and how best to get there.

That history goes as follows: All this didn't happen by accident! God had a reason for creating space and time; God wanted things to happen and in order for anything to happen there had to be space for thingness and time for happening. Who knows how many different "space/time continuums" exist, but we know that this one does and that it constitutes the matrix for all the activities in which our bodies engage. We also know that while the principles that caused all this to come into existence are consistent (otherwise things would fly apart and we'd have chaos), the activities that take place in space and time are entirely subject to change. From one minute to the next, from one breath to the next, all is change. Each of us is here temporarily and indeed, "here" is here temporarily; our planet is going to be consumed by the sun in a mere four and a half billion years (it's been here that long already; we're halfway there). The sun itself will eventually burn out, and who can say what form things will take after that? All is change.

So God created space and time. But that was not done by waving a magic wand and creating it in one stroke; it was done through a series of processes which, just as evolution has continued to do since our universe took form (and will always do), steadily followed the course that God laid out (had and has in mind). All of Creation is not static; it is constantly and eternally evolving, growing, expanding.

We know quite a lot about the history of this planet; about the formation of the solar system, the galaxy, and the Cosmos which gave birth to our planet Earth and all the life forms on it. As we trace the development of these systems within systems we find it diversifying. And by the same token we find it simplifying as we go backward in time, to where we consider it to converge in a point of origin that we currently call the Big Bang, which brought into existence (in this space/time continuum) that which became all the things we have known as Life.

But we almost always consider only material realities; the forms that things take as they manifest. Things have evolved in a kaleidoscopic series of events each of which built upon what had gone before, according to the principles that were built into the Plan (and again, without which principles we could only have chaos) and the sum of which is more marvelous than words have the power to tell. But behind all those material forms are the spiritual realities that cause all forms to manifest, of which realities we are almost entirely ignorant, and which we now need to understand.

Early (very early) in the doings of the human creature there was a massive problem, which was entirely of our doing. In order to understand how this came about we have to go back to the concept of our existing as a thought in the mind of God, the essential point being that we existed (lived) before we came into bodies in space/time on Earth (as we will continue to exist [live] after we have laid these bodies down). The idea here is that it was then, at that point in our history that we made the enormous mistake that the Bible describes in its great myths (which are both allegorical and historical) of the Fall of Man, the Garden of Eden, and the Prodigal Son. That mistake was as follows:

God created us and sent us forth into space/time with the following set of instructions: "You will find, as your various planetary systems evolve from mineral to vegetable to animal to human to what comes next, that you are 'made in the image of God', which means that you possess the same power to create reality (within the world in which you perceive yourselves to exist) that I have. You can do anything you can imagine (even though you see yourselves as just lumps of matter; of primordial mud, you can drive a car on the moon!) but there is one central, absolute, one hundred percent inviolable requirement: everything you do (everything) must be good. If you do anything (anything) that is not good (which means bad, wrong, evil, false, etc.) you will prick the balloon of perfect Life and you will then find yourself living with a mixture of Good and Evil rather than only with perfect Good."

This is what those great myths are telling us. The "Garden of Eden" is the symbolic depiction of perfect Life, from which we departed when we disobeyed that single, essential warning; when Man "fell", the Prodigal Son "left home", and we began our journey into this world of wrongness and pain.

Several things are important to observe in this connection. First, God did not "kick us out" of the Garden; we left of our own free will. Up to that point everything was perfect; nothing was wrong. God did not create Evil; we did. This is vital to understand; God is Love, and love cannot punish or in any other way be or do anything bad, wrong, evil. God is Life and Wrongness (Evil, Bad) is Death; for God to have any of these qualities, even to the slightest degree, would be to work against the very Life that God is. It makes no sense for God to work against God. God was perfect, God is still perfect, and God will always be perfect, and therein lies quite a conundrum for earthlings.

If God created everything and if Evil exists, then God must have created Evil. No, not at all; God created us and we created Evil, which is nothing more nor less than going against God's will; against God's perfect Plan, by doing anything Wrong. What, exactly, is "Wrong"? I can define it no better than I have above, but words do not nearly suffice because (as we're seeing constantly now) the meaning of words can be misconstrued, either through lack of understanding or deliberately, and right words can be used for very wrong purposes.

Our difficulty at this juncture in our history could be summed up as an inability to tell the difference between Right and Wrong.

This most essential activity can only be done with what we call our "heart"; it is something that we feel in our hearts, not something that we suspect or hypothesize, and certainly not something that we ascertain according to our emotional predispositions, upon which the forces of Evil play so continually and so successfully. You can't prove scientifically that Love is Right and Hate is Wrong but when our hearts are not broken, such things are self-evident.

So we "fell". Exactly as we were warned would happen, we found ourselves living a life in which Right and Wrong were constantly and ubiquitously mixed, which meant a world in which pain exists. In a perfect world there could be no pain because there would be no reason for it to exist; pain is God's "NO!" sign, and thank God for it because without it we could at least theoretically continue to live wrongly forever. But God built in pain; anything we do wrong causes pain. Not as punishment from God (God is Love) but as a signal to us that something is wrong, so that we will follow where the pain beckons us to its source and correct the wrong, which will then mean that the pain must cease because its cause no longer exists. What is bad will feel bad; that is what pain is. And when we insist on continuing to do anything wrong, especially when we do it with increasing power, the pain associated with that error must increase, which means that our salvation is assured eventually when the pain of wrongdoing becomes unbearable, which is what is threatening to happen now.

Can't we see the message here? Ever since we made the Great Mistake against which we were warned, we have been living with a mixture of Good and Evil and over time we have become so accustomed to this that we can't even imagine living only with Good. The analogy of Plato's Cave is perfect; we're all living in the darkness of the cave. Then one day someone is playing around with seams in the rocks and he tugs on an edge with the result that a slab shifts and a ray of light streams in, whereupon two things happen: 1) the other occupants of the cave scream at him "Shut that thing right now or we'll crucify you!!!" 2) The guy who cracked the rock realizes that there is an "outside" out there and that it's Light, all around. Whether or not he can generate any interest on the part of his fellow cave-dwellers is another matter, but he knows, and he will never be the same again. He has seen the light, which is spiritual truth; the real truth about the Reality within which we exist in the darkness of our ignorance.

Obviously, Good exists in us as well as Evil; if it didn't, we couldn't survive. But what we fail to realize is what Good and Evil, Right and Wrong really are and that acceptance of Evil has always been not a necessity (much as it seems to be simply because of force of habit) but a choice, ever since we chose (God knows why) to disobey God's warning. What we face now, with the accumulated consequences of having done so much so wrong for so long with exponentially increasing power, is the necessity for deciding once and for all (finally, thank God!!!) which master we will serve, realizing at last that to serve one is Life and the other is Death.

If we can make that choice, we will set ourselves on the path to the perfect Life that is and has always been our destiny. We are not doomed to live in this prison (cave) of Wrongness and pain forever; only so long as we believe that we have to; that this is the only possible way to live. And it is absolutely certain that at some point we will realize that we can be free, and that the way we become free is to make that Choice, between Life and Death.

Romans 12:21
"Do not be overcome by evil, but overcome evil with good." (NIV)

Isaiah 5:20
"Woe to those who call evil good and good evil, who put darkness for light and light for darkness, who put bitter for sweet and sweet for bitter!" (NIV)

Proverbs 17:15
"He who justifies the wicked and he who condemns the righteous are both alike an abomination unto the Lord." (ESV)

Romans 12:9
"Let love be genuine. Abhor what is evil; hold fast to what is good." (ESV)

Proverbs 4:27
"Do not swerve to the right or to the left; turn your foot away from evil." (ESV)

John: 1:5
"The light shines in the darkness, and the darkness has not overcome it." (NIV)

Let's head into "God (the Creator) is all power" territory; large things dwell there. Let's go back to the beginning, when God took the first steps in creating everything that exists. First we must be perfectly clear that everything means <u>everything</u>; at that point there can have been no separate thing of any kind. All there was, was God.

Very well; what did God have at hand to make anything out of? Not nothing (you can't make anything out of nothing), but no thing. God couldn't go down to Home Depot for boards and nails because there were no such things. The only possible conclusion we can reach is that God made everything out of God (since that's all there was), and this opens fascinating lines of thought.

First, we must realize that God is infinite, eternal, and REAL (that in fact, God is Life). The trick is to wrap our head around the idea that anything can be real that isn't "thingy", which we generally define as "perceptible to the five physical senses". We're so locked in to the material aspects of Reality that they seem to us to be the most fundamental reality, when in fact they are the least substantial reality. Take a close look at this sense of "substantiality" that we instinctively feel with regard to physical objects (including our bodies). First and foremost, they are solid objects. The baseball, the bat, and my body that swings the bat are three separate objects, subject to all the laws of nature that we have, over time, learned so much about. But when we go deeper we find that in a more, not less real sense they are not at all separate objects, nor are they truly solid, and their non-separateness and non-solidity become evident when we contemplate what our science has revealed about all matter; about any and every seemingly solid object.

None of them; no form of matter, is actually solid. Certainly they appear to our senses to be but that sense of solidity is an illusion. True, bang your illusory solid fist into an illusory solid brick wall and you'll have an illusory broken fist. All that is perfectly real; it is the world we live in, but it is illusory or if you prefer, a "surface" reality. Take any solid object and look at it more deeply through a mental (or physical) microscope. What appears to be any singular object is revealed to be the farthest thing imaginable from a single entity; it's actually composed of a staggering number of relatively extremely small (but still apparently solid) objects called "molecules", which themselves turn out to be anything but singular and solid; they too are actually universes made up of "atoms". And that's anything but the end of it; the atoms are made up of sub-atomic particles, which are made of strings, then quarks, and then language fails. In fact, (check it out) if you removed all the space in the atoms of our bodies the entire human race could fit in a sugar cube! Quantum mechanics then takes over, with postulations (such as things not forming until you observe them) that border on the mystical. So much for the notion that our bodies are the core of our being!

I will coin here (I think) the Principle of the More Inclusive Truth, e.g. that matter is solid and matter is not solid, but an understanding of the latter necessarily includes an understanding of the former, while the opposite is not the case. Both statements are true but the fact that matter is not solid is the larger, more inclusive truth.

With regard to the illusory nature of the solidity of matter, we have to ask what matter really is. I can think of no better term than "a dreaming", or "phenomena of consciousness", and this understanding enables us to take a different and very much deeper look at things than we generally do when we see ourselves just as solid bodies in a world of solid objects. For one thing, it solves the riddle of death.

DEATH

I N ORDER FOR us to form any meaningful concept of the human journey it is absolutely essential to understand that we do not die. It is generally conceded that one of the most common human emotions is fear, and at the top of the list is fear of death. We fear death primarily because it appears to be an end to life; so far as we can tell (because we take ourselves to be just our bodies) we become nothing when we die. We hope for continuing life but few of us have discovered facts that unequivocally or even substantially indicate that that hope is realistic, rather than just wishful thinking. Hopefully this is about to change, and change radically.

There are several problems with our normal ignorant and fearful view of death that become apparent when we put our minds to the matter; first, the idea that we become nothing is manifestly impossible since we are something (even though we see ourselves only as our bodies) and something can never become nothing. Theoretically, anything could become anything else, but no thing can become nothing. Then, so what if we were to "become nothing"? We wouldn't exist so we wouldn't care about anything, so why get our knickers in a twist over that one?

But here's the thing; we do not become nothing. Even our bodies don't become nothing; their components break down according to natural mechanisms and are redistributed into the material world. But what we most need to realize is that we are not our bodies; they are ours. We refer to them that way constantly without paying attention to what that means; what is the nature of the "owner" of our body; the "proprietor"; the "me" that experiences and uses that body. The components of our body are meaningful to us as they are assembled into the form that we "inhabit" and "use"; when they disperse we think that we're done for because we think they're us, but such is not the case.

This can only be felt; as with all things spiritual, intellectual argument will lead our horse to water but it cannot drink, which is to say that it cannot grasp the reality that mental images (ideas) represent. Ideas are like photographs of things; they are not the things themselves. Obviously, for example, the idea of Love is quite different from Love, however excellent the representation may be; Love, like all spiritual realities, must be felt. It is with that same ability to feel that we can see ourselves for what we really are; the "me" that experiences my body in the material world during the day, the exact same "me" that experiences a different body at night in the "dream world" and, can we not see, the same "me" that will experience whatever comes in the "next world". If we are to continue to exist after we leave this world, there must be something that continues; that is here and turns out to be there too, and this sense of our real self that is not its vehicle is what we could call that connecting link, although that is far too narrow a term since what it points to is the very essence of our being.

What we need, and we now need it desperately, is to realize that we can look inside ourselves as well as outside; that we can discover and thus bring to a conscious level profoundly important truths that have hitherto remained unconscious. We can see how we actually do behave, understand why we behave that way, and learn how to change that behavior in every way that it is wrong, i.e. that it should be changed. The mystics of the world have always understood this, and numbers of us awakening to this realization was a major component of the magical Sixties, with its increasing awareness of spiritual realities, some (some!) genuine gurus, and psychedelics to toss you into your inner ocean.

But again, this must be felt and there are various ways to develop the ability to do that. Perhaps the most common is meditation, where you quiet your body (easy to do), quiet your mind (very difficult until you get the hang of it), and then pay attention to what you have left, which is you. This is the real you that will never die; which will (if we are to believe the experts) continue into an unlimited future of such wonders as we cannot yet imagine, when we have solved the enormous problem that we came here to work out.

Another interesting way to look at our fear of death is to realize that it's perfectly natural, given our vital misunderstanding of what we really are and in fact, of everything else. We just don't realize what things are; that they are solid only in how they appear to us through our physical senses; that everything had to come out of, to have been formed by and literally of, That Which Created All Things, i.e. God. That all things are the original One Reality caused to manifest in different forms but always composed of the same single essence, which can only be God; the One becoming (being) the many.

At some level, deep down inside ourselves, we know perfectly well that our common sense of death is absolutely false, but since we scarcely ever look inside ourselves with any depth we're trapped in our critically limited view of this very important matter. We live on the surface of things, in the material world that is the manifestation of the spiritual realities that lie deeper; that gave (give) rise to the material world. But since deep down inside ourselves we really do know the truth, we're in a bind; we live with the constant tension that that dilemma necessarily produces. There's a lovely story about a child who had just learned what death was supposed to be and who burst out crying. His father said "Son, I was here when you arrived in this world and you reacted just as you're doing now to the idea of leaving it. Hadn't you better make up your mind?"

The only death there is, is Falsehood. Truth is Life: we must turn toward Truth. It will come to meet us, to the exact extent that we come to meet it, and it will draw us to itself until at last we are perfect. That is our goal, and that is our destiny.

NEAR-DEATH EXPERIENCES

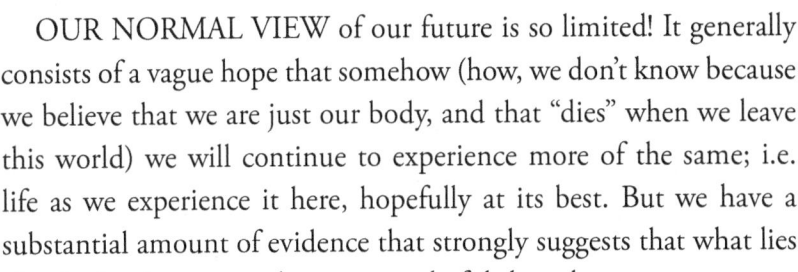

OUR NORMAL VIEW of our future is so limited! It generally consists of a vague hope that somehow (how, we don't know because we believe that we are just our body, and that "dies" when we leave this world) we will continue to experience more of the same; i.e. life as we experience it here, hopefully at its best. But we have a substantial amount of evidence that strongly suggests that what lies ahead of us is very much more wonderful than that.

We have a sizable body of literature that tells of the experiences that quite a lot of us have had when we almost died but came back here instead. Not the least fascinating aspect of these tales is their similarity; in such claims as their insistence that what they experienced as they, you might say, looked into the next world were real, not "imaginary"; that they were wonderful and in keeping with the teachings of our great mystics, who have always tried to educate us in the matter of death (and life). They accord with the experiences of people who have "gone out of the body" but were not dying, as with psychedelics or anesthetics or by other, more spontaneous means. One of the most striking of these similarities is that as people move away from this world they commonly lose all fear of death, the most spectacular example being the common agreement that at the very door to the next world is a great white light, which is the very light of Love. And this is the point where many people agree that they didn't want to come back.

Such experiences are frequently dismissed as fantasy; in technical terms, as products of our brains. But cases exist where people have had NDEs filled with highly cognitive experiences when their brain was flat-lined; simply not operating cognitively. This is the same way of looking at things that maintains that we are our body and nothing more; that every "inner" experience we have, including our basic awareness of anything (our consciousness) is a product of our brain. That is what locks us into the material world and keeps us unable to perceive the inner world of spiritual reality.

Anyone who wants to explore this vitally important question; to understand what death really is and isn't, would do well to take the matter up with someone who has had a near-death experience or other out-of-body experience. Unfortunately, far too often their accounts are dismissed out of hand, as though they were saying that they just returned from, say, Tahiti and were trying to tell us how wonderful it was, only to have us tell them that there's no such place as Tahiti; that they're just imagining it. But let's not forget; one of the things people often insist on is how very real these experiences were; it's up to us to make of that what we can.

It's difficult to exaggerate the importance of our coming to realize that our real life is not limited to the brief flicker of living in a body on Planet Earth. As soon as we become aware of what we really are, we can then open to an understanding of what the human journey really is; that it is infinite and eternal, and that we each are a living part of it. What vistas this opens for us!

THE PROBLEM

WHAT WE ARE facing now could not be simpler; it is the fact that we really have, and have always had only one Problem, which is that we have any problems at all. Every problem is defined as "something wrong"; it is an example of the real Problem, which is Wrongness itself. Of course we must deal with individual problems but we must do that by putting them all in their proper context; as examples of the real Problem; Wrongness. As we do this we begin to realize that "practice makes perfect"; yes, it is often hard in this wicked world of grey to separate black from white (Wrong from Right) but every shade of grey is made up of just black and white, and when we sincerely try to separate Right from Wrong we find that doors open that could not have opened until we turned toward them rather than away from them. "Try me, and see if I will not open the windows of Heaven and pour forth such a flood that there will not be room to receive it."

What lies ahead for us when we wake up, see what the Problem really is, and deal with it as it should be dealt with, is nothing less than our release from all suffering of any kind, which must result from our removing the cause of all pain, which is Wrongdoing (including wrong seeing). This is our destiny; when and how we achieve it is up to us. To date, we have been trying to serve two masters, the results of which are making it perfectly clear that this cannot work in the long run, although it may have appeared to work for a time (debts can be allowed to build but eventually they must be paid and in this case they draw compound interest). God has so arranged things that we cannot live wrongly and get right results; if we persist in wrongdoing the results must at some point build to where they overwhelm us, which is exactly what is happening now.

So what will we do? Will we continue to lock ourselves into seeing only the material world in which our bodies exist? Will we continue to ignore the lessons of history with regard to what we have done before and the results that it produced? Will we just allow our mistakes (fear, hate, doubt, ignorance) to continue to worsen? We can do that; indeed, if we look at the surface of things it's hard not to conclude that that is exactly what we are doing.

BUT… if we look beneath the surface we get a very different picture of what is (still) possible. We are not over the Cliff yet, even though it seems that with it so close and with our pedal still near the metal in top gear headed straight for it we're bound to go over. Indeed, we may, but still we may not; if we orient our eyes, minds, and hearts sincerely toward Truth, all the powers of Heaven and Earth will come to our assistance but the only way we will be able to see what that means is to "Try me, and see…" Granted, it would seem that only a miracle could save us now but consider; a miracle is not something that can't happen. It's something that did happen that we didn't understand until later. If we turn toward Truth the Cosmos will open and miracles will pour down in Martin Luther King's "mighty stream". Let us not forget; Life is entirely miraculous; wholly magical. A single blade of grass is a symphony of miracles.

Spiritual seekers have for a long time been fond of the "Great Invocation", which goes as follows:

THE GREAT INVOCATION

"From the point of Light within the Mind of God, let light stream forth into the minds of men. Let light descend on Earth.

From the point of Love within the Heart of God let love stream forth into the hearts of men. May Christ return to Earth.

From the center where the Will of God is known let purpose guide the little wills of men; the purpose which the Masters know and serve.

From the center which we call the race of
men, let the Plan of Love and
Light work out; may it seal the door where evil dwells.

Let Light, and Love, and Power restore the Plan on Earth."

Let Walt Whitman state the case: "I say that no man has ever yet been half devout enough, none has worship'd or adored half enough; none has begun to think how divine he himself is, and how certain the future is."

CONCLUSION

SO WHAT IT comes down to is that 1) we made a terrible mistake when we turned away from cooperating with God in the never-ending and wholly good process of creation. We went off on our own, separating ourselves from God (Reality) and creating our own world filled with false illusory realities. 2) over time we became so accustomed to living with the mixture of Good and Evil that we couldn't even imagine living only with Good, thus trapping ourselves in this prison of Wrongness and pain by our own hand. 3) As they had to eventually, the consequences of all this wrongdoing have built to the point where they now threaten to overwhelm us. 4) the peril of this existentially critical moment is immense but the promise is infinitely greater if we can see that the peril is entirely the result of the Wrongness that we have accepted and that therefore all we have to do to reverse our course is to stop doing bad (wrong) things. 5) that this utterly simple thing is, in this wicked world, often extremely difficult to see and do, is irrelevant; we can do this and we must do this. Difficulty has never stopped us when we're clear about our goal, and that goal could not be simpler: "Be ye perfect..."

It's just that simple. Truly wonderful things are just waiting for us to open our inner doors to them; allowing the bubbles of perception in which we each live to expand, rather than fearfully clinging to them, defending them against what we so often take to be our enemy but which is actually our savior; the real Truth of things. It's a pity that we made it necessary for us to be shaken awake rather than awakening in the natural course of evolution, but that was one of the inevitable results of our Great Mistake.

We now have not just the opportunity, but the necessity of correcting that mistake; if we can see that and act upon that understanding, not only can we still escape The Cliff that looms so close; we can enter a whole new life that our experts have always told us only awaited our recognition of it to manifest as the magnificent reality that has always been our destiny. When and how we achieve that destiny has always been up to us; now we can take a quantum leap in that direction if we can understand and make The Choice with which we are being so starkly presented. Which master will we serve?

"Let Light (Truth) and Love (the root motive behind all of Creation) and Power (our will, hand in hand with God's will) restore the Plan on Earth."

We have a wonderful opportunity here to move ahead with God in the never-ending process of evolution, becoming free at last from Wrongness and pain forever. Shouldn't we seize that opportunity with both hands?

ADDENDA: FOOD FOR THOUGHT

W hen we take a real interest in things, things turn out to be really interesting. In fact, fascinating; here is a scattering of what to me at least, seem to be very interesting things well worth thinking about.

FAITH

WHAT WE NEED to do now is to ask ourselves in every situation "Is this right? Is this according to God's will? Should I (we) be doing this? Rather than operating from fear that our needs will not be met unless we commit our usual wrongs (political or economic competition, war, racism, etc.), we must have faith that things will always work out best for us to the exact extent that we truly worship Right and strive to put that attitude into practice in our lives. Only when we consciously and deliberately do this can we realize that real faith yields real results; that it is not fantasy or wishful thinking but tapping into the deeper spiritual realities that in fact inform all surface (material) realities. It's not so much a question of praying for a car and getting a car as it is assuming that with a certain attitude things will simply work out for the best. Often things will not work out the way we might have ordered them off the menu but over time we will find that what arrives on our plate actually meets our real needs better than what we would have ordered. But in order to judge accurately how well or poorly things are working out in any given moment we need to understand by what standard we should be judging, and that in turn requires that we know what we are really here for; where we're actually going. Only then can we stand back and see clearly how things are going, not in the usual context of our short-sighted material view of things but in terms of our real journey.

What we are here for is to play our part in God's Plan, understanding that that Plan is an endless unfolding of God's Great Dreaming, as Lao Tzu said, "from wonder into wonder". Playing our part means turning wholly towards Right (God's will); if and when we do this, wonderful things happen ("Try me..."). But before we can be properly set on that course we must free ourselves from what is holding us back; our Problem, which is acceptance of Wrongness in any form, ever, holding us in the prison of error and pain that we must now realize is only as permanent as we suppose it to be. It is up to us to arise as truly human beings and make the ultimate choice between Good and Evil, which choice will free us in a way for which our innermost hearts have always yearned but which our minds have been unable to grasp as being real. Up until now, evolution on this planet has proceeded generically as it worked to produce a vehicle for a creature that was able to understand these things. It has done so (something worth contemplating deeply in order to begin to understand its importance). That wonderful creature is us, and now further evolution depends upon our exercising that ability to see what is Wrong and to make it Right, thus freeing ourselves and the whole of Creation to advance into a whole new octave of being; as Shakespeare said, "a consummation devoutly to be wished".

In the last analysis it is our hearts that perceive what is Right (True, Real, Good). It is important to realize that our hearts are not just "connected with" the heart of God; they are one with it. They are individualized centers of Love, which is universal and infinite; it is at the very core of all things. And as with a hologram, each individual "piece" contains the whole within itself, as the whole contains the piece within itself. This is one of the most profound mysteries; it is what Jesus was referring to when he said "The Kingdom of Heaven is within you" even as you are within the Kingdom of Heaven.

HABIT

W HY CAN'T OUR minds accept "Salvation", "Heaven", "Freedom", "Perfection", as real, not just imaginary impossibilities? This is a fascinating question that goes deep; by way of an answer I would suggest that it is the power of habit, in this case deriving from the fact that time and time again, day in and day out, throughout what we're aware of as human history our hopes have so often been dashed. This is such an integral part of our life that we accept it, in both good and bad ways. The good way is "sucking it up" and accepting that it's almost inconceivable that anyone will entirely escape pain in this life. This we do with great strength and courage, but limited wisdom. The bad way; the great mistake that we couple with this resignation and strength, is to assume that this condition is a necessary part of life; that it could not be otherwise.

This is utterly without rational justification, any more than the idea that we couldn't fly just because we never had; this attitude is a habit that we badly need to break. What makes us feel this way about what is and is not possible? I suspect that it is the result of our having had our expectations shattered, our dreams crushed, so many times and in so many ways that each of us has a limit to what we're willing to dream lest too high a dream be dashed, which would be too much to bear. So to dream of the ultimate good, of absolute and permanent perfection, is totally out of hand; altogether too much to hope for as a reality or to even allow ourselves to imagine. To bare our hearts so completely; to believe in something so utterly and permanently wonderful but then to have that ultimate hope crushed would be unbearable, so we just don't dare to dream that high.

Bad habits (particularly the habit of having them) tend to build; with each bad habit we accept we increase the likelihood of our accepting more and worse habits. How far will we go ("Once you accept anything wrong you're in mortal danger of accepting <u>anything</u> wrong")? Habit is like a millstone; a bad habit rolls along behind us, threatening to crush us if we don't keep running away from it. But by the same token a good habit rolls along in front of us, pulling us ahead rather than threatening to crush us. Either way, habit has momentum that builds as the habit is continued.

One of our most unfortunate habits is limiting our dreams as discussed above; it is certainly understandable but it keeps us seeing the future as having terribly limited possibilities, the best of which, perfect Life, we insist is wholly out of the question. But unless we can see anything as a real possibility, how can we play our part in bringing it about? And really, don't we all wish that things were all right? How many times a day do we say "All right!"? Deep inside ourselves we know, we know.

In the hectic rush of modern "civilization" we have separated ourselves from our past; from the precious knowledge that has been carefully preserved by indigenous societies who are still connected to our Mother Earth; to the natural systems that we call "the environment" of which, regardless of our forgetting, we are a part and upon which we are wholly dependent.

"White man clever but not wise." Far from having understood, respected, and protected our planet's natural systems, we so-called "civilized" folk have declared open war on them (with remarkable candor and ignorance we call it our outright "war on nature") and we have come perilously close to "winning" that war; a Pyhrric victory if ever there was one, as is becoming undeniable.

Not only have we largely destroyed our planet rather than protecting and preserving it, we have viciously attacked the indigenous people who resisted our rapacious genocidal advance and we continue to attack anyone who attempts to point out the tremendous error of such ways. "Environmentalists" (i.e. sane people who see what's happening, who are alarmed and who attempt to call general attention to how critical is the situation into which we have allowed ourselves to drift), protesters against wrongdoing of all sorts, "whistleblowers", anyone who is attempting to make things right, are viciously opposed by the powers that be, who are firmly in the grip of Mammon. This is not just murderous; it is suicidal. It is wrong, it is against God's will, and we should not be doing it.

IDEALISM

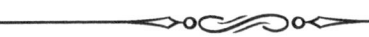

BECAUSE OF THIS tragic habit of limiting our aspirations it takes great courage to be an idealist but it shouldn't. As with so much else in this flawed world, we've got this wrong. Far too often we see idealism as wooly-headed, rose-colored, wishful thinking that is of no use, even counterproductive, because we take it to be simply unrealistic. The idea here, when we think it through, is that there is a limit to how good things can be, and right there is the fatal flaw in our so-called "realism". The idea that anything is "too good to be true" is nonsense; within the limits of Cosmic Law (e.g. two and two will always equal four) anything is possible and nothing is too good to be true. Far from being unrealistic, (quote me on this!) idealism is the highest form of realism.

It works like this: first we form the "ideal"; our mental image of what would be not just a good goal but the perfect goal; the "ideal solution" to our situation, in practical terms the best outcome we can imagine. We take this first step completely free from any considerations of whether or not this solution is "feasible", "realistic", or "possible"; thinking only about what would be the most desirable outcome. Then the next step is to consider whether or not this outcome is actually possible, realistic, and feasible. This is where boldness is called for, viz. the Wright brothers, Einstein,

Galileo, or any of our great "inventors" who simply refused to take no for an answer. One assumes that the goal is possible and says "OK. There is no good reason not to believe that this can be done. Now, working from what we have already done that we know is possible and knowing how to do it, what steps would lead us from here to where we want to go?" This is where faith, imagination, intuition, and inspiration come marvelously into play; if what we were hoping for is actually impossible; if it goes against God's rules, i.e. "natural law" (e.g. if it would involve two and two equaling five) then it will quickly become apparent that we've hit a dead end and that's that. But if it is possible, these channels within us will open (because we turned toward them rather than away from them) and "from wonder into wonder existence opens".

MANIFESTATION

THEN COMES THE question of how things move out (evolve) from their source in that which is permanent into space/time, where all is motion and change. This is an extremely recondite subject at which I can only hint but obviously an important one, and it is most useful to have at least some sense of how Creation works, in order to appreciate our place in the overall scheme of things.

A rough outline of the process might be put this way: before any separate thing appears there must be some "substance" out of which it is made. So to put it in the simplest terms, God created a "layer" consisting of the basic substance out of which all (separate) things would be composed. But this "layer" was (is) composed of thought, which was then caused to become space, time, and all things within it. Anyone interested in looking more deeply into this process can investigate "Animus Dei"; the mind of God, which created (creates) and then acted (acts) upon "Anima Mundi"; the mind of the Cosmos, to produce all of Creation.

THE DEVIL AND GOD

ONE OF THE most unfortunate aspects of our normal view of things is our idea that The Devil is real; an actual entity that is not only real but so powerful that he is capable of challenging God; indeed, of fighting him to a standstill. So back and forth they go, with the good things in life alternating with the bad things as God and The Devil contend with each other in their attempt to influence our behavior.

There is a tremendous misunderstanding here. The Devil does not exist except as we create him with our imagination, which is perhaps the most consequential example of our ability to create illusory reality that while false nonetheless appears perfectly real to us. What The Devil (aka Satan) really is, is a symbol. An enormously important symbol, representing the sum of everything wrong in life, but only a symbol nonetheless. We imagine a host of minor devils, all in the service of the Big Guy and they too are symbols; what each of them represents is one or another of our common serious human problems, each of which is (as discussed above) no more nor less than an example of the Big Problem, which is Wrongness itself; more exactly, our acceptance of Wrong in any form, ever, as a necessary part of life.

God, on the other hand, is real. Not just real but the only reality. Everything that exists is a form of God manifesting as that thing (thought forms; aspects of The Dream). Everything that we create (including our illusions) is no more nor less than us using our creative power ("made in the image of God") to shape God into forms of our choosing but at base they are all forms of God. There is nothing but God.

We cannot fully understand God; the part cannot encompass the whole. But there are several fundamental aspects of God that we can understand; they may seem difficult to grasp at first but they are well worth the effort. To wit: 1) Being everything, everywhere, God must be infinite. Point your finger in any direction and go. Don't stop until you realize that if you do, something must continue to exist beyond that point. There can be no point where Reality (God) stops, therefore it is infinite. Our minds cannot grasp what that is but they can grasp that it must be the case. 2) God must be singular; one Ultimate Reality, because if there were two (or more) they couldn't be infinite; they would be separate and by definition limited realities (this is the real meaning of monotheism; that there is only one God [one Reality], within which exist all subsidiary realities [including all the gods and devils we have invented]). (3) God must be alive because how could the source of all life not be? 4) God is Love. This cannot be demonstrated intellectually; it must be felt by our hearts; which God gave us precisely for that purpose; to feel love.

Most unfortunately, we with our creative power create for ourselves false realities and The Devil is the sum of all those illusions. The whole story of God fighting The Devil is no more nor less than the story of the Great Human Mistake that we made when we "fell" and the strife commenced between Good (God) and Evil (The Devil) in which we have been engaged ever since. But from which it is our destiny to emerge with God the victor, with all that that implies. Meanwhile, we continue to dream up and live with false realities that constitute "the thousand ills that mortal flesh is heir to".

This process and its effects are demonstrated with marvelous clarity by what we call "hypnotism", a procedure in which another person's will is substituted for our own with our consent, and an illusory reality is skillfully suggested to us (precisely the process by which our perceptions are constantly influenced by, for example, our media, our advertising industry, and our politicians) with the result that our illusions become real to us.

I watched a spectacular display of this process when I was touring the country with a musical trio; at a hotel gig one night there was a hypnotist on the bill. He hypnotized our lead singer and gave him the suggestion that he could see through the clothes of all the women in the audience, and it became immediately obvious that he could. To him every one of them was naked. The point is that for him, what he was seeing was just as real as anything the rest of us were seeing. Back we go to the difference between "real Reality" and "illusory reality"; our task now is to learn to tell the difference between these realities and to wholly orient ourselves toward real Reality, which is Truth, which is Right, which is Good, which is the fundamental quality of God's Plan and of our destiny.

With so many of us currently willing to accept such huge, important, patently obvious falsehoods as Donald Trump having won the 2020 election, willing to support on such utterly fraudulent grounds the dismantling of our democratic system and to deny and ignore the other enormous dangers that we all now face, could it be any clearer how important it is for us to consciously and deliberately make this fundamental choice between Truth (Right) and Falsehood (Wrong)?

DESIRE

ONE OF THE greatest of our religions is Buddhism, which has demonstrated over time that it has much to offer, especially in helping us achieve peace of mind in the face of the world's troubles. But there is an enormous flaw in how it is too often understood, as holding that all suffering is the result of desire and that therefore the cessation of suffering can only be achieved by extinguishing desire, thus achieving perfect balance and peace. There are several things seriously wrong with this view of things; first, we understandably and correctly desire to eliminate suffering. This puts us in a bind similar to the one we meet when we claim that "nothing is certain", to which the obvious reply is "Are you sure?" So on the face of it, the idea that desire is intrinsically wrong cannot stand the test of basic logic; it is inherently self-contradictory.

But more significantly, that idea fails to recognize the vital role that desire plays in everything that is deliberately done, ever, anywhere. God desired that things would happen, so God created Creation. We, "made in the image of God", meaning "endowed with the same ability to create reality for ourselves", were told "Do whatever you can conceive of to do (but every bit of it must be good)." Which obviously implies that we would want to do things, which is desire. To eliminate desire would be to eliminate the very movement, the activity that is conscious life and progress, because without it there would be no motive for action. Theoretically it would produce peace, but it would be lifeless.

The problem is not desire itself, which is the sine qua non of all deliberate activity; it is desiring wrong things, plain and simple. The Dalai Lama recently said this clearly; it is not desire per se that is the problem. It is desiring things that we should not desire; in other words, wrong things, and back we go to the basic focus of this dissertation: Right vs. Wrong.

UNITY/SEPARATION

EVERYTHING IS CONNECTED; everything. On the surface of Life things appear to exist separately from each other and there is nothing intrinsically wrong with this; it's how things work (how could I pass you the butter if it didn't appear to us to be separate from the bread?). But it is an illusion at base; literally, nothing is separate from any other thing, and this goes for our internal as well as our external reality; when we search for the line that actually separates one object or one person from another we cannot find it because it doesn't exist.

No thing has an "edge"; what appears to be a real singular thing dissolves upon close examination into molecules, then atoms, then sub-atomic particles and so on until we see that all sense of solidity disappears, so there can be no edge in reality, although the illusion of one is perfectly real and useful. We perceive ourselves fundamentally as our bodies; separate objects among the vast number of other objects in the material world, and internally we also separate ourselves in accordance with that vision of ourselves; of what we are. This results in our day-to-day conscious life consisting almost entirely of personal concerns; food, clothing, shelter, emotional needs, etc. As a practical matter we generally consider everything else in the world largely in terms of how the world can serve our personal needs; not in terms of how we can serve the world, and this too often results in a lack of concern for the needs of other people and of all other life forms on our planet (the "environment"), of which we each are as much a dependent part as our finger is a part of our body.

We need to see ourselves as servants of Life; that is the best (really, the only) way to get Life to serve us as it was designed to, and this means that we should have faith in God not just to ensure that our personal needs are met but to make of our selves channels through which the needs of all the rest of Reality can be met.

REALITY/ILLUSION

TO THE EXTENT that we withdraw from Reality we create illusion, the supreme illusion being that there is no such thing as Reality. This is a trap that we fall into all too often; we say "You can't be certain of anything." Indeed, we can question anything; we can insist that two and two may not equal four, that our space missions could have been done in movie studios, etc., and this is generally rational. If we decide to check something out and we discover pertinent facts, we can question any of them. If a person appears to be a solid source of information we can question them; there is no point at which we can say that anything is certain unless and until we use our God-given faculties of body, mind, and heart as they were intended to be used and trust them to show us what is real. After all, everything is real; it's just a question of what form of reality a given thing actually is and how to position ourselves so that we see it as it is instead of as it is not.

Back we go to The Problem; before the Fall, in the state that is symbolized as the Garden of Eden, there was no illusion, and when we free ourselves from this Problem there will again be no illusion. Things are what they are (what else could they be?); when we achieve "salvation" we will see them exactly as they are and our entire living will be in accord with Reality, no longer polluted by illusion, i.e. Wrongness.

We are at a critical point where substantial numbers of us have become so deluded that we actually believe that there is no such thing as Reality or Truth; that everything is uncertain and one person's opinion is as good as another's. We are now on a mass scale thus unable to tell truth from falsehood, with the result that enormously harmful falsehoods are being accepted by large numbers of us; no better example of this regrettable fact exists than the citizens of the United States of America letting such a stunningly ignorant and harmful person as Donald Trump con them into allowing him to take over the Presidency.

The recent election, while decisive (and thoroughly proven to be so) was by no means a "landslide". Eighty-one million of us voted for Biden but seventy-four million voted for Trump, even after four years of his madness and mayhem (which his hypnotized followers see as wisdom and success). Those of us who did this and who continue to accept as real all manner of things that simply are not real, e.g. the notion that Trump actually won the election, are behaving (in this regard) insanely. Believing things to be real that are not real, especially when the facts are right in your face, is defined as "insanity"; what we have here is a wave of mass insanity. Really. Literally. So what is going on?

The division between those who insist on truth; on ascertaining facts and dealing rationally with them, and those who are ruled primarily by their emotions, not by a desire to see the truth of things, has grown to where it is one of the most threatening problems facing us. We need to understand what lies at the root of this important difference; why so many of us are behaving this way, and the only way we will be able to do this is to listen to each other without fear or prejudice, so that we can understand what actually underlies these belief systems. We must somehow see how to keep each other in our hearts as God's children; every single one of us members of the human family. We can and should "hate the deed" if it is wrong, but if we lose our love for the doer of the deed we lose ourselves.

When we listen openly to supporters of Donald Trump and of the Republican Party that has accepted and supported his way of doing things, we find (among far lesser motives) perfectly understandable reasons for their discontent, together with a regrettable ignorance regarding what has been responsible for those conditions. Basically, what we see is their dissatisfaction with the way things are going in terms of our national affairs, specifically in how they are affected as individuals. If you see yourself and/or your friends losing their jobs, especially if those jobs are taken by people of a different skin color who you may have already been taught not to love and who may be willing to work for less money, it is understandable for you to demonize those people. And if your government appears to you to be allowing "those people" to flood into the country when they are (you feel) such a threat, then it is understandable that you will be in favor of tightening restrictions on immigration. And if your President (playing you like a fish on the hook) tells you "Right you are! These so-called immigrants are thieves, rapists, etc. and we've got to keep them out!!!" he will appeal to your darkest emotions, and to the extent that you are accustomed to allowing those lower angels to dominate your perceptions you will likely agree with him and support his proposals, regardless of the harm that they cause to your brothers and sisters in the human family, each of whom has needs that are just as real and just as legitimate as your own.

And as the nation becomes more and more the melting pot in which we used to take such pride, with people of all nationalities blending in one inclusive society, the greater seems the threat to the "way of life" to which we are clinging when we cannot see beyond the threat that we believe this poses to our personal wellbeing. This attitude is understandable but it is based, yet again, on illusion far more than on reality.

One of the most significant misunderstandings underlying these views is ignorance concerning the proper functions of government. The idea here (put into high gear by Reagan and flogged by Republican "leaders" ever since) is that each of us is and should be on our own. The ideal is Horatio Alger; the person who through hard work and perseverance lifts himself (and now herself, to a degree; misogyny is still as prevalent in our society as racism) up by their bootstraps, achieving and enjoying the "American Dream" of owning a home, being fully fed and clothed, plus being able to afford the boat, sports car, or whatever allows you to have fun on weekends. For most of us this is not only not easy; for shockingly large numbers of us it is impossible. That being the case, it's easy to sympathize with the feeling that any government program that takes any of our (tax) money and gives it to anyone else is robbing us of our hard-earned money. It's the old "Why should I support some deadbeat?!"

Again, this attitude is understandable but it is terribly ignorant and heartless. Go out into the street and talk with some of the "deadbeats" who are "loafing around doing nothing", and you will quickly realize that the bootstraps argument just doesn't work, because these people have no bootstraps. If their skin happens to be black or brown or red or yellow; if they happen to be female or differently gendered, if they happen to be physically or mentally "challenged" then not only will they be likely to have no bootstraps; they are likely to have had any such tools deliberately withheld from them. And a great many of them did have bootstraps. They did pull themselves up and then what? The economy took a dive (not their fault) and they lost their job. And their home. And God knows what else. How broken does our heart have to be to not understand that any government worthy of the name must do everything in its power to help these people, not scorn them, and then do everything in its power to alleviate the conditions that struck these people down?

This wretched business of "taking my money to support deadbeats" is so wrong! The basic principle of collecting taxes from everyone is sound, practically and morally. When everyone chips in a modest amount the big ticket items can be paid for; essential public needs (such as infrastructure, health, etc.) can be met with minimum expense on the part of the individual. No better example exists than Medicare For All, which is classic insurance in which everyone puts in a little and those who need it take out a lot, with nobody begrudging them that assistance because "there but for the grace of God go I" and we are our brother's keeper.

GOD AND MAMMON

WHY DON'T WE see this? Why is there such strong resistance to such obviously good ways of doing things? Fear, plain and simple; fear of not having enough of something. Having to meet our needs "by the sweat of our brow", rather than trusting that we will have everything we need to do what we should be doing. Without that faith that "God will provide" (just as God provided us with this amazing body and this wonderful place to put it) we see only the outer aspect of all things; we see our supply as limited, with everyone in competition with each other for whatever they need and want. A "zero-sum game" with limited resources, in which when one wins another must lose. This entrapment in the material view; this preoccupation with self-interest, is what we symbolize with "Mammon" (in effect, one of The Devil's top lieutenants). It leads us to focus on the things of this outer material world rather than on the inner spiritual realities that gave (give) rise to all of this world, including our bodies, minds, and souls, with their particular needs. That which formed all things can certainly sustain all things, and it will in the normal functioning of Creation unless it is prevented from doing so, which is exactly what we have done ever since we turned away from God, the inner, to the material world; the outer.

It is as if we'd pulled our plug out of the wall and then forgot that there was a socket.

The practical result of this turning away from the deepest realities leaves us perceiving only the shallowest ones. Rather than happily making ourselves servants of Right (God), secure in the knowledge that we need not fear, we make ourselves servants of Mammon. Fear drives us into the arms of Mammon; Faith drives us into the arms of God. And the voices of Fear are so loud ("If it bleeds, it leads")! It's hard to tune our personal radio away from the loudest stations to the quiet ones that have the most value but that is what we must learn to do, with our material as well as our mental media. Back we go to the fact that if we keep doing things wrong with increasing power, sooner or later things must go critical, which is what they're doing now. But again I emphasize the immense promise of this moment, as the wonderful poet Amanda Gorman said at Biden's inauguration, "if only we are brave enough to see it; if only we are brave enough to be it."

YOU ARE BEAUTIFUL

D O YOU HAVE any idea how wonderful you are? Each and every one of us is a being that we call human; the blooming (budding, actually; we have yet to bloom) of the magnificent evolutionary flower that is a vital part of God's Creation, which is God's Plan in action. Any way you look at it, so long as you really look, we are marvelous! Just take our bodies; when we see what they actually are; the intricate complexity of their parts, their systems, and the harmony of their functions, we are amazed. Then, fascinatingly connected with our bodies are our minds, which are also wondrously complex. The body is the machinery; the mind, whether in its unconscious or its conscious modes, controls the body. And beyond ("behind") our mind lives our soul; our living conscious real self. Each of us is so very wondrously made!

And do we have any understanding; any deep feeling of how wonderful our planet is? We've had a pretty good look out into the universe now in all directions and what do we see? Not so much as a blade of grass, not a bug, not a tree, not a flower, not an animal. Rocks and gases seem to be about it. Of course it's possible that anything you can imagine is out there but so far we see not one shred of evidence that there is anything anywhere that can remotely compare with the variety, the wonder, and the beauty of the life forms on this gorgeous planet, which is a reality, not a possibility. How can it be that instead of treasuring this inexpressibly wonderful gift, we're trashing it? And now we think it's a good idea to take the very creatures that are doing that and send them to Mars, where they'll supposedly flourish and save mankind? The mind reels. If we can't get our act together here with the magic cornucopia of gorgeous gifts that this planet affords us, how will we do so on the hostile rock of Mars? I see nothing intrinsically wrong with going to Mars or anywhere else but to put any significant amount of resources into doing that at a time when our resources are so desperately needed here on Earth seems misguided, to say the least. Go to Mars if you want to but only after we've got our affairs in order here.

And as we have already discussed, we should be constantly aware that all these systems everywhere are composed of matter that at base is simply not solid. Cells, atoms, all of it is fundamentally forms of consciousness, and the question is "what causes them to take that form?" Something certainly does, and that question is well worth pondering. We desperately need now to dig deeper; to look beneath the surface of the material world into the spiritual world behind it. The importance and the beauty of these spiritual realities cannot be overstated.

PROTO-HUMAN/FULLY HUMAN

B EFORE WE CAN truly call ourselves human we must allow what we call our human qualities to predominate, not our animal qualities (in the sense of our continuing to allow our animal nature as opposed to our human nature to control us when we don't have to while the other animals do). These human qualities are what Abraham Lincoln called our better angels, first and foremost among which are our hearts. In order to become fully human, we must at all times use our hearts as the primary arbiter regarding everything we think, feel, and do. Then we must use our minds as they too were intended to be used, intelligently ("sapiens"); to shape our thinking according to what is Right, which is determined by our hearts. Then, we must use our bodies also as they were intended to be used; to put our desires into action, making life what it was (is) supposed to be. This is exactly what we were told to do in the Garden of Eden; whatever we wished to do, so long as every bit of it was Right, i.e. Good. This is the goal; to have life be entirely good. We need to gird up our loins, set perfection as the only sensible goal, and allow our ability to tell Truth from Falsehood to grow, as it surely will if and when we sincerely want it to.

Genuinely wanting to know the Truth of things is the key. So much of the time we form our sense of things according to our emotional needs, thus making that sense of things real to us. When we do this we are not acting primarily from a central motive of learning Truth; we are attempting to satisfy emotional needs which all too often lead us away from rather than toward Truth.

In that state we use our minds not to explain the truth of how things really are but to explain it away. An agile mind can explain anything away; using a little imagination I can rationally claim that the moon is in fact made of green cheese; that all arguments to the contrary are simply false. I know it's really made of cheese because (my hypnotist) said so, and explained it fully. When it was claimed that we had surveyed the moon for years and found not a shred of evidence of cheese, my "leader" pointed out that this was totally false; a scheme concocted by the "fake media" to mislead the public. Likewise, when it was pointed out that we had gone to the moon and brought back rocks with, again, no evidence of cheese he countered that this too was completely fake, all done with concocted videos in the NASA studios (all those "astronauts" in the media? Actors. Rocks? New Mexico); that the whole idea that the moon is not made of green cheese was a deliberate, very clever hoax designed to convince us of things that simply were not true, so as to maintain control over our minds. The truth is that all those lies about the moon are intended to cover up the fact that Kraft is secretly working with NASA as we speak, arranging to mine that cheese ahead of the competition that will surely emerge when the truth comes out (remember Chris Kraft, NASA's first "Flight Director"? Eh? Eh?! What does that tell you?). Those of us who realize this feel sorry for the rest of you who have been so successfully duped into believing that there's no green cheese on the moon.

All this raises the question of how a human being is to reliably, actually tell the difference between what is really real (true) and what is really illusion (false). Any evidence that is brought forth for any claim, true or false, can be questioned indefinitely; you tell me your source and I'll say it's false. Show me the source for your source and I'll say that's false, and so it goes; no matter what you show me I can question it. So how are we to reliably arrive at the truth of things? The key is to want to arrive at the truth of things. When we sincerely want to know the truth, doors open that could not so long as we lacked that basic desire; when we lacked the courage to question things openly regardless of our emotional predispositions, sincerely desiring to know the Truth wherever it might lead, which is to say unafraid of learning where we have been wrong; indeed, anxious to know if we have been. Things then "occur" to us; things come to us through avenues of which we had not been previously aware and all manner of things just seem to happen that make things clear to us. We begin to realize what "the ring of Truth" means. And we begin to understand why Lao Tzu said "From wonder into wonder existence opens." It's so important to understand that as we shed Wrongness we open to Rightness, which is wonderful, beautiful, infinite, and permanent. "Rightness" simply means "Reality"; real reality, not illusory reality.

Our world today is hugely astray; terribly far from the Truth of things and we're suffering the consequences, which have reached a critical point. Can we not see that we're in an existential level of trouble, that it's all due to the wrong things we've done and continue to do, and that all we have to do is to stop doing bad things? Nobody said that's easy to do, because of the strength of habit, but it could not be simpler.

I would suggest a few basic guidelines for a personal checklist to judge how effectively you are actually seeking truth:

1. Do you truly want to know the truth, whatever it turns out to be? This is the most essential requirement; it opens the most important doors to learning.

2. Are you angry? Overly skeptical, overly optimistic or pessimistic? Are you emotionally clinging to any particular view? Any of these things will distort your vision, like a pebble thrown into the lake that must be perfectly calm to truly reflect the sky above.

3. Are you listening? Openly, unafraid, being objective in your assessment of what you're hearing? The blind men and the elephant are the perfect analogy here; if they listen to each other they can learn from each other, but too often they just turn fearfully away from each other because they can't handle a different opinion; they can't see it as being as real as their own, just from a different point of view. Zero-sum thinking; if I'm right then anyone with a different view must be wrong. Most unfortunate.

4. Are you willing to be shown that you're wrong if you actually are? How we do resist being corrected when in fact that is one of the best things that could happen to us, expanding our awareness of how things really are!

5. Can you see when you've reached the limits of your understanding? Can you be comfortable with that uncertainty?

The first Commandment: "Thou shalt love the Lord (Truth) with all thy heart and with all thy soul and with all thy mind…" Matthew 22: 37-38 "This is the first and great commandment". Truth is Life; Life is God. As we turn deliberately toward Truth it turns toward us. Is it really such a stretch to suggest that this is the essential journey we should be on?

BTW, I much prefer to think of the Ten Commandments as the Ten Promises. Nobody has ever had much success commanding me to do anything, whether they were right or wrong, and methinks that most of us feel that way. And this is not necessarily a bad thing; not by any means because when all is said and done we are each responsible for our own life. Nobody's hand but our own should touch the rudder of our personal boat; certainly we should be open to advice from any source regarding any aspect of sailing but each of us must be ultimately responsible for assessing the value of any advice and translating it into what we do at the helm of our own ship. This, not incidentally, is the core principle that has made the United States of America great in the eyes of the world: freedom of the individual, not just to make a buck (although that, too) but to be absolutely on one's own two feet, judging for one's self all matters of importance in one's life.

So when the Commandments say "Thou shalt…" I choose to interpret "shall" as "will" in the future tense: "You will (sooner or later) love the Lord thy God with all thy heart, and with all thy soul, and with all thy mind"; you will "love thy neighbor as thyself", and so on. Promises of the glory and wonder of the Life we will all live when we solve our Problem and live rightly in the real world.

"BY THE SWEAT OF OUR BROW"

ONE OF THE most fundamental characteristics of evolution is the increase in individual and group intelligence displayed by succeeding life forms. A plant is more intelligent than a rock; an animal is more intelligent than a plant, and a human is more intelligent (as far as we know) than any other life form on this planet (although the cetaceans [whales and dolphins] are highly conscious creatures; how highly conscious we don't yet know). While it is true that intelligence is intelligence (the activities of trees, for example, exhibit amazingly complex intelligence), as evolution has proceeded each major category of individual life forms (mineral, vegetable, animal, human,???) has exponentially expanded the scope of its intelligent activities.

We humans have demonstrated a wonderfully high degree of intelligence, from our material inventions to our scientific discoveries to our philosophical realizations. How then can it be that with all this intelligence we have brought ourselves to this critical state of affairs?

We have made it perfectly clear that we have the ability to save ourselves; to save the world, to understand what we are doing and why we are doing it. So why aren't we saving the world? Why are we destroying it instead? No rock, plant, or other animal is going to save the world; it's up to us. We've tied the knots; we must untie them.

We aren't doing it not because we lack the ability but because we lack the will. The reason why we lack the will is because we're just too busy doing what we do, to which in our ignorance we see no alternative or (even under our present conditions) reason to seek one. We don't have faith; we don't understand what faith is; how essential an ingredient it is in right living. We're so busy trying to swim that we can't see that if we would just relax, we could float.

MASS INSANITY IN THE UNITED STATES

FOR THE PAST half century, the world has been watching the United States of America go mad. Which is to say, to turn strongly (even further) from what is right, good, and real to what is not. The old adage about boiling the frog certainly applies (except that no frog would allow itself to be boiled; it would jump out of the water); it would seem that we will accept (get used to) almost anything, provided that it happens by degrees. Our government has steadily devolved from an instrument of service to public needs (which is of course its purpose and which to a significant degree it was when Roosevelt turned it in that direction in the Great Depression), into an instrument of open and notorious service to a small group of individuals who seek power and money above all other considerations. Particularly since Ronald Reagan flung open the doors to this way of doing things, with the likes of Lee Atwater, Karl Rove, and Newt Gingrich saying "We make the rules now. We decide what is going to happen. The rest of you talk about it but we do it. We act and you react." So intense and exponentially rapid has been this concentration of power that we now find ourselves in a situation where a staggering, obscene amount of resources is in the hands of a few individuals, while huge numbers of American citizens are "food insecure" (what a horrible euphemism for "don't have enough to eat"!) and actually dying in our streets. And such is these spiritual infants' level of social and general awareness that the richest of them (who has more money than the GDP of a hundred and forty of the world's hundred and ninety-five countries!!!) says that he can think of nothing better to do with his money than to put it not into immediate desperate human needs but into colonizing space! The second richest person agrees, believing that the way to save the world (which he believes is in critical, maybe hopeless condition) is to take the very creatures who are ruining this magnificent planet and transport them to the barren rock of Mars, where they will presumably flourish and preserve the future of the human race.

This is madness. And it is matched by the insanity that now controls our Congress, which has degenerated to the point where the Republican Party (which represents a substantial minority, not at all a majority of American citizens but which has gathered a vastly inordinate amount of power through the anti-democratic Electoral College [Donald Trump lost the popular vote in 2016 by three million votes, which according to the most fundamental rule of democracy means that he lost the election], as well as gerrymandering and various other schemes that keep people who oppose them from voting) has such a vise grip on affairs that they ruinously follow in the most craven manner imaginable, by far the worst "President" that this nation has ever allowed to take power. And when enough of the citizenry (and the Democrats) woke up to the point where they impeached him (twice), the Republicans refused to convict him, even when he incited a mob assault on the Capitol with the intent of staying in power by disrupting the certification of the election in order to overthrow the legitimate government by force. This is an insurrection against the government of the United States, but it is much larger than that. These people believe that the legitimate government is actually illegitimate; that

Donald Trump won the election, which was "stolen" from him. The facts are perfectly clear; Biden won the election, absolutely without question. But horrifyingly large numbers of us shut those facts out, choosing to believe (and thus make real for themselves) something that simply does not comport with the reality in the world around them. The dictionary defines this as insanity; "a state of mind which prevents normal perception".

Here we see massive reversal of the truth, e.g. regarding which groups are actually responsible for the vast majority of violent acts (we're being asked to believe that the attack on the Capitol was actually a clever "false flag" operation by the commie Antifa [anti-fascist] leftists; an attempt to discredit the real Americans, i.e. the Proud Boys, the Republicans, etc.). We see denial of the stark reality of climate change. We see this ghastly pandemic described as a hoax. All of which is vigorously supported by "Christian fundamentalists", regardless of the obvious fact that they are behaving directly contrary to the teachings of Jesus Christ. So what we have here would appear to be one third or more of the American population literally going insane. Why? What is really going on here?

This is not easy to understand but like everything else we do, we do this for a reason, whether consciously or unconsciously and as always, if we bother to look closely certain things become clear. First, this insanity is the result, at base, of fear. Deep, unreasoning fear. Generally, understandable fear in the face of the "slings and arrows of outrageous fortune" that batter us daily with circumstances that range from mere nuisances to unbearably painful events. At the present time we are confronted with a storm of painful circumstances on every level, from individual desperation at the loss of our job, our home, even quite possibly our life (say, from COVID 19) to a generally mostly inchoate sense of doom from several developments that hover around the edges of our awareness, first and foremost among which is climate change, which threatens global catastrophe right at our doorstep.

When we're afraid we tend to react emotionally, to the extreme detriment of our ability to see what we're dealing with, to think clearly about it, and thus to act effectively to deal with it and remove the cause of our fear. So it controls us, rather than having us control it. The result is a vicious circle in which the very things we fear are worsened by our inability to deal with them because of that fear.

Among the most powerful influences that perpetuate and strengthen this vicious circle are what we call our "right wing" media (with huge irony because they're so terribly wrong), who are extensive and tremendously influential in putting false images in the minds of so many of us who are so easily misled. We must remember; to those who accept these lies they are not lies. They are reality; just as real to them as the real reality is to those who see it instead of the falsehood.

So we have shockingly large numbers of people believing things that are not true and acting on the basis of these beliefs, which are constantly stoked by corporate media whose chief concern is the money they can make from these ruinous activities ("It may not be good for America but it sure is good for us!"), combined with our Congress, who are fundamentally hand in hand with the media although there are heroes in Congress who see through the smokescreen and do their best to counter the lies and fix the enormous problems to which those lies give rise.

But in the end we have to come back to the fact that "the only sin is ignorance". Ignorance breeds fear; fear breeds anger, suspicion, hatred and separation in the human family, all of which we're seeing now on every hand. It is perfectly understandable that so many of us are disillusioned by our government, which has been not just abandoning us but working directly against us for at least forty years, serving the interests of the corporate oligarchy and the Military/Industrial/Congressional Complex rather than the interests of "we the people". But so many of us are so ignorant of the real causes of that regrettable state of affairs; of who is actually responsible, how, and why. So in our ignorance we allow ourselves to accept such ruinous ideas as that our government can only be our enemy, just because it happens to be acting that way. Thus allowing ourselves to forget the obvious fact that the whole purpose of government is to serve the interests of the many, not the few, and that if we would just wake up and insist on it that could be made to happen.

But far beyond the effect of our ignorance on our national affairs, in a much broader sense it shows that we have no understanding of the real journey that we're on; of its nature, its direction, or its eventual outcome. We see every form of trouble coming at us with or without warning but we recoil from most of it in ignorant fear, being unable to deal with it. But if we knew the wonderful truth we could not possibly act this way; we would realize that our troubles are temporary, that they have a cause that can be dealt with, and that that cause is single, not multiple. It is simply our acceptance of Wrong as a necessary part of life when it is not, whether in our personal, national, or human affairs. We must turn from illusion to reality; we must become sane.

IGNORANCE IS NO EXCUSE

CTUALLY, IT'S AN excellent excuse; how are we to be expected to obey a law if we don't know it exists? The proper meaning of the statement is clear enough; ignorance of a law doesn't keep that law from operating; you snooze, you lose. The problem is ignorance; simply not knowing the rules or not caring; deliberately breaking a law, which is in itself generally ignorant, particularly when we're talking about natural law (which unlike human law is perfect). Ever since we left The Garden we have broken God's natural laws both ignorantly and deliberately. We constantly do both as a matter of course.

What should we expect of ourselves? In our fearful and punitive legal system, we expect everyone to know and obey every law; "ignorance is no excuse". Our judges and juries do mitigate to a meaningful extent but an enormous number of people have been sent to jail who never should have been because of this hard-hearted attitude (the same one that expects people to pull themselves up by their bootstraps when they have none).

Freedom doesn't mean that we don't have to obey any rules. "Original Sin" was and is the notion (the feeling) that we have the power to do anything we want, so anything we choose to do is OK. And we do indeed have the ability to do anything we want (anything that is actually possible) but the cause of The Fall was our insistence that just because we were free to do whatever we pleased it didn't matter what we chose to do. We were warned; we were told that it mattered (matters) very much indeed. That there was one central rule that lay at the base of all rules; of all the intricate operations of Creation, every one of which was called into existence by and continues to be maintained by Love. If we did anything (anything) that was not infused with Love, that was (is) Wrong. When we depart from Love we depart from Reality and enter into illusory reality, in which we've had one foot ever since The Fall. "Right" means "in accordance with God's will"; "Wrong" means "against God's will". God's will is Love.

One of the strangest things I can ever remember doing was when I was maybe six or seven years old. My mother was ironing; the hot iron was sitting on the ironing board and for some reason (why?) I wanted to touch the iron; to check it out for myself. My mother told me that that would be a bad idea; that I'd get hurt, but rather than just saying okay and not touching the iron, I said "But I want to!" (WHY?!). Again my mother warned me, and again I said I wanted to. I think she gave me a third strike, and again I insisted; I touched the iron, burned my finger, and then rather than taking responsibility for my action I blamed my mother for letting me do it! "When I was a child..."

What on earth did I think I was doing? I think I was repeating Original Sin; I felt my freedom to do what I wanted and nobody was going to tell me what to do or not do. My personal desire was all that counted; I had a lot to learn. But there's an easy way and a hard way; I (and we) chose to learn the hard way. I really don't understand why I so foolishly insisted on ignoring Mummy's loving warning but then, neither do I understand our having so foolishly ignored God's loving warning. I can just hear God saying "Don't touch hot irons. Yes, you can, but you shouldn't. You will, anyway? OK; go ahead. See you later." Meditate on those last three words and weep.

Freedom must be exercised with responsibility; responsibility to the ultimate law of Love. Like me with the hot iron, we take any suggestion that we shouldn't be doing anything that we happen to want to do (e.g. not wearing a mask during a pandemic) to be an abridgment of our personal freedom but it is not; the coin of Freedom has two sides that cannot be separated without consequences. We have the power to separate them (to ignore one side) but we do not have the power to avoid the consequences, which at this astonishing time are coming home to roost all around us. We do whatever we want to do just because we want to and we can, regardless of natural law. If we loved ourselves, if we loved all of God's creatures, if we loved God, we would not behave this way.

We practice a wretched calculus wherein we know perfectly well that we're doing something wrong but we want to do it, so we do. Because it was wrong it feels wrong, which is to say, it hurts, either immediately or later as its consequences become apparent. But rather than realizing that the purpose of the pain is to lead us to its source, correct the wrong, and thus remove the cause of the pain, we make three enormous mistakes. 1) we say that God is punishing us for our sin, which is completely incorrect; God is Love and love cannot punish, our pain is the direct result of our wrongdoing. 2) We make a deal; we say "I can handle that level of pain and the pleasure I get from this act is greater than the pain, so I'll keep on doing what I'm doing, and I'll take the pain simply as the cost of doing business. That's life." Thus completely misunderstanding the purpose of pain, which is to lead us out of darkness into light. 3) we (try to) make another deal with the Cosmos: "I can live with this view of life; don't make me change it!" We cling to our world view because we're afraid to let it expand naturally.

So we keep ourselves in Plato's Cave by our own hand. We keep on doing things that we know are wrong, figuring that the cost is acceptable. Then we make another mistake; we fail to realize that the same wrong if repeated with increasing power must yield increasingly severe consequences. We have been doing so much so wrong for so long and now with such great power that those consequences threaten our very existence on this planet. Really; it's that serious.

RACISM, PREJUDICE. ETC.

THROUGHOUT HUMAN HISTORY we have exhibited an unfortunate habit of looking down on each other, with an amazing array of excuses for doing so. One suspects that if not the original impulse, certainly one of the most powerful was "the other"; the enemy who just as with other animals was constantly at the borders of your group's chosen territory if not actually at your cave door. In the same way that we continue to demonize those who we feel are threatening us ("Japs", "Gooks", "Ragheads", etc.), painting them in the colors of what we take to be their worst features with little or no recognition of their best ones (no general can allow his troops to think about the enemy's good features; it plays hell with their will to fight), we surely demonized those who threatened us back then. To say that we looked down on them is putting it mildly.

But we go much further than that; we so often look down not just on other nations but also on people in our own nation with different skin color, religious or political beliefs or, it would seem, just about any other difference that we can find between us not just nationally but in our communities and even our families. We can't use war as an excuse in such cases, although far too often fear of some sort of attack, often imaginary, fuels the concern we feel regarding the target group. So why do we do it? What is it in us that causes us to behave this way?

It seems clear that any attempt to put ourselves in a superior position to another person has to stem from inadequate self-esteem. This shows in our competitive habits; whether in the sports arena (military fly-overs and the national anthem ["the bombs bursting in air, the rockets' red glare"] at football games are no accident), in politics, in business and so many other areas, the "winner" is the one who counts; Top Dog ("Top Gun"), who is elevated in the eyes of his/her admirers and consequently in their own. Second place barely counts, even though the runner-up may have displayed almost exactly the same prowess as the winner. This is one of the reasons why power corrupts; when everyone is "looking up" to you, bowing and scraping, begging for autographs or whatever, it's hard not to get a swelled head, which you would have no use for if you could realize how wonderful you and they are just as we all came from the hand of God.

If only we could realize what it means that "we are all equal in the eyes of God"! That each and every one of us is a precious masterpiece, brought into existence through inexpressibly wonderful processes that must have themselves been authored by a Creative Force of, let's just say, the highest value. We must realize that while we are the tip of the spear of evolution on Earth; the creature that is the end result of all that has gone before, evolution is not finished (nor is there any reason to believe that it will ever be). We are still becoming; growing into what we will be. But now as already noted, we must want to grow; we must awaken to what we really are and what we should be doing here. When we do come to understand these things a whole new world will open to us but as with Plato's Cave, one of the first things we must realize is that we are (self-) imprisoned in darkness, ignorant and fearful of the Truth (the Light). We must come to see that the light of Truth is the exit from the Cave, and that it is the only way out ("the way, the truth, and the light").

Can we not yet see that the millstone around our necks; the Curse that causes all curses and that we brought upon ourselves in The Fall, is no more nor less than Wrongness in any form and to any degree? This is now our task; to realize that we made and continue to make a terrible mistake that has haunted us ever since with the "thousand ills that mortal flesh is heir to", then to see that if we can deal with this central disease all its symptoms (individual problems) must disappear, thus freeing us to live life as purely good with no evil at all ever again. This is the choice that now faces us. Of course it has always faced us but we have turned away from it; now the consequences of that turning away are cascading down upon us. That is the cancer that has caused every sore on the body public or indeed, in the inner and outer life of each and every one of us. If a thing (an idea, an act, anything) is wrong it must carry with it God's announcement that it is wrong, i.e. pain of some kind; in other words, it will feel wrong. Of course we should deal with every problem on its own terms but most important is that we realize that every problem is caused by our central disease; failure to realize that anything wrong is wrong. Then we can cure every problem (yes, that is the promise) by curing the disease. Would that not be, in Shakespeare's words, "a consummation devoutly to be wished"? It all comes down to The Choice. The conscious decision to turn away from everything that is wrong, realizing that the difference between Right and Wrong (through however dark a glass we may be able yet to see it) is not at all a matter of opinion but of Cosmic Law, built into all of Creation by That Which Created It.

THE BIBLE

WE SHOULD REALIZE that the Bible, like many other documents produced by human hands (historical ones in particular), is replete with errors, misinterpretations, and contradictions. Nonetheless, buried like a great underground vein of gold is the profound spiritual understanding that some humans have always managed to acquire and have attempted to set forth for whatever benefit the rest of us can get from it. The book is centered on the teachings of Jesus Christ (although it sets forth the spiritual awakening of the human race as traced from Abraham through The Prophets and Moses to Jesus), which when properly understood are seen to be unequalled in the depth and value of their truth. So great is this value that even though few of us yet see below the surface, vast numbers of us intuitively realize that the teachings do indeed have great value. Not for nothing is The Bible the best-selling book in human history.

One of the most important of the great Biblical promises deals with the very serious question "All evil gone forever?! If it happened once it can happen again; what assurance can we have that it won't?" Aside from the fact that when we've really learned a lesson we have no need to repeat it; we learned what not to do, the Bible states that "I will remember thy sins no more", which means that there will be complete forgiveness; no grudges held, no punishment, nothing like that at all; not even the memory of the wrongness. It will be as though the Great Mistake had never happened. We would do well to dwell on what a glorious vista that puts before us.

The key to this entire business is our turning away from Wrong in any form, toward Right; the results of doing so will launch us into a future that is infinitely larger and more wonderful than anything we yet seem to be capable of "realistically" imagining. All we have to do is to solve the one Great Problem of Wrongness in order to escape from this prison that we have accepted for so long as the only possible way to live. This is and has aways been our primary task here on Earth; to work this Problem out. It is our destiny to do so but it is up to us to do it.

"THE CLOUD OF UNKNOWING"

"THE ONLY SIN is ignorance." There is a tremendous barrier or veil that forms the walls of our prison; that encloses us in our bubble of ignorance. When we fell, the very act of choosing to do what we did, to turn away from God, clouded our ability to see that Truth and Falsehood are not equally necessary aspects of Life. All we've been able to see ever since The Fall is a mixture of Truth and Falsehood, and we take this mixture to be an essential, unavoidable part of Life. But as conditions continue to worsen on our planet, more and more of us are willing to contemplate things that until these conditions forced us to, seemed impossible. Things like not having war, everyone on Earth being fed, clothed, and sheltered, protecting rather than destroying the environment; that kind of thing. We are beginning to ask why the resources available to humanity cannot and should not be shared effectively throughout the human race (the fundamental principle of Communism and Socialism). All manner of things that we've never devoted much thought or action to in the past because we were just too busy with our own affairs, "earning our living by the sweat of our brow", are demanding our attention. Now the question that obviously must be asked is "How's that (our normal behavior) working for us?" The answer is clear. It's not working at all; it's led us to the brink of global catastrophe.

But something very much larger is afoot than just having things run smoothly on this planet; the planet itself is going to be consumed by the sun eventually; our systems of normal life won't count for much then. However, we will continue to count for very much indeed; what we really are will continue. But the task at hand at this critical point in our journey is to discover just that; what we really are, and to make all our plans in the context of that realization. The concerns of one material lifetime in a body can no longer preoccupy us; we must now seriously consider the great life questions because they all have answers, and in our currently perilous situation we desperately need those answers.

The truth is, God has a Plan. All this did not just happen for no reason. That Plan is more wonderful than words can tell. All that we see around (and within) us is that Plan unfolding, and it has now unfolded to a very important point where through its evolutionary processes it has produced a creature that is beginning to do what God had in mind as the culmination of this stage of the Plan; to understand the Plan and its own vital part in it.

But our "vital part" involves a level of freedom to create that must be accompanied by a commensurate level of responsibility to use that freedom rightly, not wrongly. And right there is the Problem; the single source of all our specific problems, each and every one of which is just an example of the central Problem: acceptance of Wrong in any form whatever, to any degree. Jesus knew what he was talking about when he said "Be ye perfect, even as your Father in Heaven is perfect." We are currently making it clear what must happen when we use our creative power irresponsibly; sooner or later the consequences of this behavior must "go critical", as it is doing now and as we were warned by our Creator that it would if we didn't wake up, see the truth of the wrongness of Wrong, and set ourselves the obvious goal of eliminating it entirely.

Just think of it! What could life be if there was nothing wrong (which means with no possibility of anything being wrong ever again)?! We don't allow ourselves to raise our sights that high; to dream of something so wonderful, and that is understandable, given the vicious circle in which we trap ourselves.

But brothers and sisters, the real truth is that this Perfect Life is the only real life there is; the illusory reality that we build around ourselves notwithstanding. Those illusions (falsehoods) are the only problem we've got; they all stem from our acceptance of Wrong as a necessary part of life. Our goal must be to free ourselves from this prison of illusion and pain, and it is our destiny to do this. We can do this and we will do this, one way or another. God's Plan continues to unfold; the human creature is budding and soon to blossom into an understanding and a freedom to create properly (rightly) that we can scarcely imagine. What lies ahead of us when we see, accept, and obey it is nothing less than what each of us in our deepest heart has always yearned for; it is Perfect Life. To find, to live that Perfect Life we must seek it with the understanding that it is our natural goal and our destiny.

Realizing this enables us to deliberately make of ourselves fitting instruments for the forward motion of God's Plan. Indeed, the continuation of that forward motion is now dependent upon our awakening, ridding ourselves of the millstone of Wrong that we placed around our necks in The Fall, and advancing into doings of pure Love and Light.

All we have to do is to stop doing bad things. Period. Pay no attention to the constant repetition all around us of the notion that this can't (even shouldn't!) be done; that it's "unrealistic". It is that notion and only that notion that makes it effectively unrealistic. Not only can it be done; it must be done. And regardless of the seemingly insurmountable odds against our awakening to these realities in the evidently short time we would currently seem to have left to operate on this planet, that very peril should stimulate us to take things very much more seriously than we are accustomed to doing. That kind of serious thinking seems tedious until we actually practice it; then it becomes increasingly less of a burden and more of a delight. I'm reminded of a song that John Sebastian wrote about a dream he had in which "all of the heavies were light as a feather". When we live in an awareness of the real truth of things; of God's Plan and our part in it, we live in joy, beauty, inspiration, and lightness of spirit.

COSMIC LAW

G OD'S CREATION HAS rules; inbuilt consistencies without which none of it could hold together for an instant. If there was no such order there could only be chaos, in which nothing could happen because there would be no way for it to happen; nothing could hold together at all for any length of time. So God has rules; certain basic principles which obtain throughout all of Creation. These rules appear to be few in number; they apply throughout all the different layers (or "planes") of Creation. For example, the law of momentum and inertia is manifest at the material level, but also at the mental level, in the form of habit. A bad habit has momentum that can be very hard to overcome; to stop, turn around, and go in the opposite direction. And when we come to a stop, inertia can make it difficult to get started again. But when we overcome that inertia we find that a good habit has momentum that carries us along in the right direction just as a bad habit carries us in the wrong direction.

Another of these rules is that Good and Bad are not at all matters of opinion; they are cosmic law. Good must yield good results and Bad must yield bad results (even though it may seem otherwise in the short run). And perhaps the most basic rule of all: Love is at the root of all things, which is why our hearts are our most important tool; we simply cannot go wrong with Love. These laws manifest at all levels of reality; physical, mental, and spiritual. Our science has uncovered several of them in their study of the physical world; gravity, electro-magnetism, etc; they are the forces (some say angels) that keep things being what they are and doing what they do. But no force in this world is greater than the human will, guided by Love.

WHY DO WE RESIST TRUTH?

ONE OF THE most astonishing things about our behavior is the consistency and vigor with which we resist turning the handles of our doors and opening our personal bubble to a larger perception of truth. On the face of it this is insane; an expanded awareness of Reality can only be in our interest, yet we resist it fiercely. Why? Because we are terrified of being proven wrong. Again, why? If we are wrong about anything, it can only improve our life to see that and to expand our understanding, but we resist that opening with incredible insistence. What are we so afraid of?

One of our most fundamental needs is our craving for certainty, which is not just legitimate, but an absolutely vital need. But the only real certainty there ever could be is living awareness of Truth; real Truth; standing on the rock that will never be pulled out from under our feet. They don't call it "The Fall" for nothing; the instant we went against God's will we lost the certainty of Truth because we departed from it, and thus pulled the rug (the ground) out from under ourselves. Not enjoying the sense of falling (uncertainty) and needing to justify our doings to ourselves, we created false ground to jam under our feet and halt our fall. But that falsehood had to become apparent sooner or later, and at this critical juncture in human history it's doing so in a very big way. The false ground we stand on is being pulled out from under our feet; we're falling precipitously and the only thing that can stop it is the solid ground of Truth.

We accept the painful consequences of our illusions as a necessary part of life but then, rather than understanding the purpose of that pain, seeing its connection with our illusions and using it to free ourselves, what we tend to do is to make our deal with the Cosmos: "This is the reality that I can live with; let me hold onto it." Thus we lock in our worldview and shut down our growth. We're afraid that if we stay open to the truth it will bring us pain and God knows, that's understandable; it has bitten us over and over again in the past, but only when we were wrong. As we move toward truth, things must improve in our lives until there is no wrongness and pain left. It is our destiny to achieve that goal; to live entirely in the Real World, which is perfect.

We can tell that the indescribably wonderful Life that we've been told is our destiny really is our destiny, by applying the simple law of conservation of energy, as follows: All this did not happen by accident. God has a Plan. That Plan is perfect; we screwed it up in our neck of the woods but that situation is temporary. For it to be permanent would mean to thwart God's (perfect) Plan, which is self-evidently impossible because thwarting (utterly defeating) God's Plan would necessitate having more power than God, who is all power. Therefore, God's Plan must be fulfilled, sooner or later; it is impossible for it not to. So our destiny is assured; it is certain. But how and when we achieve the next forward step and become free is up to us.

We're fully accustomed to creating false realities that we take to be real; our lives are filled with self-created realities that are simply wrong. But we don't see what we're doing because these things seem real to us and we're clinging to these realities for our sense of security. That's why we're so defensive when any of these false realities are questioned; whenever anyone suggests that we're wrong about anything.

We need to summon the wisdom and courage to stop clinging to the false realities that had to lead eventually to the chaotic situation we're in now. We need to accept the uncertainty that we have always feared, but which is such a large part of true awareness; the more we know, the more we realize we don't know. The vast mystery of eternity and infinity will always be before us, but that in no way precludes our continually expanding awareness of Reality, which (and only which) provides the certainty; the "solid ground" that we so rightly crave. The whole universe of real reality awaits, and it is not only the solution to all our problems; it is perfect Life; the Garden of Eden. It is real; it has always been there waiting for us to return home to it and our welcome when we, the Prodigal Son, do return will be more wonderful than we can imagine.

ARE WE UNIQUE?

WHAT WITH THE infinite reality that surrounds us, who can say what other space/time arrangements there may be; what other creatures exist out ("over"? "up"? "down"? "in"?) there? What we can say is that so far as we have yet been able to see, we are not just unique but spectacularly so. Here we inhabit a planet so special that even with our current ability to examine the far reaches of our universe we have yet to discover a single planet that can even remotely compare to this one.

Words cannot encompass the variety on this planet of color (what painting could ever exceed the loveliness of a flower?) or of sound (what human endeavor could exceed the beauty of the Western Tanager singing as I write?). What words are equal to the intelligence innate in, for example, the mesmerizing regularity of a crystal, or in the simple fact that things don't fly apart; that nature is order, not chaos? What language shall we use to describe the reality of Love, of the miracle of a child being born into this world, of our capacity to wonder and appreciate, or the satisfaction of real human accomplishment? Of the inner as well as the outer quality of any and all forms of life?

Simply because of the enormous numbers of planets that we now realize are "out there" and the uniformity of natural systems that we see everywhere we look, there is a strong tendency to assume that there must be other life like us out there, by which we mean another life form or forms with which we can communicate intelligibly. This is our general assumption at this time. But consider; it is only the weight of numbers that makes us feel so sure that this must be so. We have as yet not one shred of credible evidence of the existence of any such life form anywhere but here on this magnificent planet.

And as far as the weight of numbers is concerned, consider the odds against the human creature ever having come into existence anywhere; the odds would appear to be so hugely against such an event that they're right up there with the big numbers that make us think that it must have happened elsewhere. A study was done some time ago by a scientist attempting to calculate the odds not on the human race appearing, but on the existence of just one molecule of one of the essential amino acids necessary for our arrival; he found that for that to occur according to the laws of chance in the time frame and under the conditions here where it did happen, would have required a universe sixty sextillion sextillion sextillion times the size of the Einsteinian universe! The obvious conclusion being that none of this happened by chance.

So the odds (and it's all odds; we have no evidence) against anything like us ever appearing here or anywhere would seem to be so enormous that they are comparable to the chances that there are other life forms "like us" anywhere. We may, in fact, be unique in all of Creation (it may be that the whole purpose of forming this universe was to bring forth the fully evolved human being), and this is a thought to hold because if we are or might actually be unique we should treasure ourselves and treat ourselves and our world with the loving respect that we and it deserve. And we should seriously consider our responsibilities as a vital, possibly unique part of God's Plan.

WILLINGNESS TO EXULT

IT TAKES COURAGE to be happy, because of our fear of being hurt as we have so often been (because of our Problem). Being wounded in love, for perhaps the best example, we pull back in our willingness to feel because we fear being wounded, thus making it impossible for us to really give or receive love. So courage is required; the courage to risk being hurt.

"Feeling good" means feeling goodness. The goodness is real; it is all there for us to perceive. Our usual way is to look at a flower, for example, and say "That's nice", but when we're willing and able to look more deeply at what that flower really is; what it is that is actually being that flower, we are not just mildly moved by its beauty; we are thrilled. And so it is with the whole of Reality. This is one of our most harmful habits; we shut down our capacity to feel because we fear the pain that we have every reason to expect in this world; but that feeler feels goodness as well as badness, so when we close it down we shut ourselves away from God.

SILENCE

MOST PEOPLE THINK that silence is nothing. They don't realize that it is everything. It is the voice of what everything is. Each thing has its unique voice but behind each thing; in fact, actually being each thing, is God, of whom every thing is a unique individual expression; one note in the symphony that is the Song of God. Everything that exists came out of not-thingness; every sound came out of the Silence. The Silence is not nothing; it is everything. When we realize that we can hear the Voice of the Silence.

This is difficult if not impossible to put into words; as with all spiritual realities, it must be felt. Obviously, the value of silence cannot be appreciated through our ears or by our minds but as with all things, our minds can lead our horse to water. But it is with our heart; our ability to feel, that we drink; that we perceive the deeper meaning of anything and everything.

Listening to the silence is meditation; in this frantically active world that can be difficult but practice makes perfect, and there are few more valuable activities than meditation for the human creature to engage in. Meditation is listening to God.

ABANDONING PERFECTION

WHEN WE PRICK and collapse the balloon of perfect Life we begin the process of living with Wrong. The question then is, having accepted anything wrong, where will we stop? Having allowed ourselves to be persuaded that any wrong is right, what other wrongs will we allow ourselves to be hypnotized into accepting?

With our lives now so filled with a virtually infinite variety of wrongness; with that wrongness having built to the point where we now must either correct it or face catastrophe beyond anything that we have ever seen, each of us must ask ourselves "What will I accept as a legitimate reason to do anything wrong?" The answer must be "Nothing."

This would seem to be impossible. If, for example, someone attacks me, am I to refrain from defending myself? If someone attacks a loved one (or group, or nation) what are we to do? Jesus said "Turn the other cheek", and this would seem to be the answer; if nobody would pick up a gun there could be no war. But in this world people do pick up guns and use them to harm others. Are we to just let them run loose?

If we are to turn this world into the ideal world that in our hearts we all yearn for and which we are told is our destiny, then what other way can there be but to follow Jesus's instructions? "Realistically", I can think of no better advice than to say "Do our best" to improve, in the hope and expectation of having things improve as they need to. But with time so short to extricate ourselves from the now so nearly disastrous consequences of the imperfection that we normally accept without question, it would seem that we'd better get on with it.

And rather than just trying to find ways to mop up spilt milk, shouldn't we do our best to figure out how not to spill the milk? Isn't it self-evident that it's better to prevent a problem than to have to deal with one? Well then, let's study how to prevent problems. And why not all of them? Really; fully seriously, why not? Rather than rushing about in an attempt to deal with a nearly infinite number of separate wrong things, why not put the medicine at the root of the tree and deal with Wrongness itself, thus addressing all the problems with individual leaves? How much simpler!

WHAT IS OUR GOAL?

T HE WORLD IS filled with individuals and groups of people trying with intelligence and heart to solve every problem we've got. Some of them are well known; most of them are anonymous except to their associates. Aside from the specific goals that each of them have in mind, what are our larger goals; the general hopes, even intentions we have for the whole of humanity and of all the rest of God's creatures? The truth is, we seldom think on anywhere near that expanded a scale; we are preoccupied with our immediate concerns.

But some day, in about four and a half billion years, the sun is going to consume the Earth; what then? It would seem that all our finest institutions; our greatest accomplishments, will count for nothing at that point in our journey. This of course does not mean that we should lessen any of our good efforts, but what it does mean, particularly at this extraordinary time when these institutions and accomplishments are so strongly and imminently threatened, is that we should expand our vision to look beyond that point where this material world is consumed. We should contemplate not just the individual lives on the surface of this planet, but the totality of those lives; their real nature and the journey that we're all on and that as it turns out, will not stop when the planet does.

The only way we can understand that journey; the only way to see what it is and to help it unfold, is to realize that it is a spiritual reality, not just a material one. We must come to know ourselves as that within us which does not die. When we see that, the "human journey" expands exponentially, in fact infinitely. We see that the life we are actually living here in these bodies on Earth is not limited to those bodies and this Earth, none of which is actually solid anyway. It's all a dreaming; as has been well said, we are not material beings having a spiritual experience (if we consider ourselves to be having one); we're spiritual beings having a material experience. It is absolutely essential that we look inside ourselves and realize what we really are; the dreamer, the liver of our lives. The "me" that uses "my" body. The "feeler" that perceives everything we experience. Only when we see this do we realize (make real to ourselves) what we really are; only then can we define the human journey of which we each are a part.

The stirrings of this realization arose in a fascinating way in the West (these things had been known since time out of mind in the East) when Sigmund Freud invented western psychology. He realized that we were not just our bodies; that there was an infinite ocean within. Freud, being a fairly neurotic chap, was terrified by that infinity and, having dipped his toe in, he immediately withdrew it and proceeded with what he is justly famous for; an essentially mental, curative enterprise. But Carl Jung took a greatly expanded approach. He realized that it was essential that the infinite reality within us be explored, not turned away from; that the whole cosmos of spiritual reality was there within us ("The Kingdom of Heaven is within you"). This remarkably rapid evolution of our inner exploration from a curative mental enterprise to an overtly spiritual endeavor is one of the most important developments in the history of Western thought (my great aunt Fanny Bowditch [Katz] was treated by Jung from 1913 to 1918, just as this vital change was taking place; she was wonderful and fascinating to talk with). Our scientific and technical accomplishments are marvelous indeed but it is the spiritual search that will yield lasting results of the highest significance.

PERSONAL STORIES: THINGS THAT TUGGED ME OUT OF THE BOX

I'VE HAD A number of things happen to me over the years that gave me a strong feeling that a lot more was going on than meets the eye. I think most of us have had similar experiences but because they're "out of the box" we hesitate to talk about them for fear that we'll be seen as weird rather than interesting. How much better off we'd be if we shared these experiences and mused together about their significance! Here are a few of the experiences that come to mind:

Back in the mid-Sixties when I was living just off what became the Haight-Ashbury district in San Francisco I found myself essentially broke (again). I had four dollars and I needed six to take my girl friend to an Ali Akbar Khan concert down in Big Sur. I was an impecunious singer of folk songs in my mid-twenties at the time, and the whole slew of concerns about being broke came flooding back: how was I going to get the money I needed the next day? Beyond that, what was I going to do to make money in the future? How would I ever support a wife and children? Etc, etc, etc.

As I lay in bed worrying about all that, my mother's soft voice came to my mind's ear saying "Behold the lilies of the field; they toil not, neither do they spin, yet I say unto you that Solomon in all his glory was not arrayed like one of these." This seemed relevant; I didn't think I was stupid or lazy; I just couldn't believe that there was no option but to "toil and spin". It seemed to say that if I stopped worrying and simply trusted the Cosmos (the Creator, God, whatever) to take care of me, it would. Not necessarily, I have learned, by doing anything directly for me but by giving me whatever I needed in order to do what I should be doing (just as it gave me this amazing body and this wonderful place to put it), so that if I just headed into the next day with the intention of living rightly, I'd have what I needed. "Right!" said my lower self, and we went back and forth for a while until no new arguments were forthcoming; I was just spinning my wheels and I resolved, even if just as a scientific experiment, to trust God to furnish whatever it was right for me to have the next day.

The next morning, I stepped out on the deck of the friend's apartment where I was staying and thought "Today God is going to take care of me." After a brief chat with my lower self ("Hey; we settled all that last night. Unless you have something new to offer, get thee behind me!") I glanced at the mailbox of the apartment downstairs, where I'd been staying until a couple of months before. Once in a while there still was mail in there for me, so I decided to check it. There were two letters for me, one with a check for $42.50 for a recording gig I'd played with Peter, Paul, and Mary months before in New York, and the other from an aunt, celebrating my very late uncle's birthday with a check for a hundred dollars (she'd never done that before). Understandably, I think, that was a very significant experience for me (that hundred and forty-two dollars would be equivalent to more than a thousand dollars today); I decided that this attitude of trust in God was not just realistic but of primary importance. I resolved to ride that horse as well as I could as long as it ran. It's never faltered, and my cup runneth over.

Don't get me wrong; I don't want to sound like a Pollyanna here. In my eighty-three years in this world I've made my share of mistakes and seen my share of hardship (loss of a child, cancer, serious injury, etc.) but all that pales when compared with the good things with which this life has blessed me. So I'll burn your ear all day about how real God is (the only reality in fact, as I've attempted to explain above) and how, loving all of Creation as God does, God can be trusted to provide what is needed for us to do what we should be doing. We need to be as trustworthy for God (in faith) as God is for us; then we can walk hand in hand as we were always intended to.

When we moved into the hills of Northern California in 1971, seven miles off the nearest paved road, there were only about ten people living in the roughly ten thousand acres of former sheep ranches and logged forests into which we moved. Shortly after we moved here, a neighbor who had bought the ranch house parcel in a ranch adjoining ours had a young lady visiting him who turned out to have lived in the same apartment in Cambridge, Mass. that I had lived in. Upon discovering this, she handed back to me four of my strange and wondrous books that I'd left in that apartment; what were the chances of that?!

Another event that was so unusual that it made me feel that more was going on than we could yet understand was when I, my wife, and a friend were headed north to Eureka and pulled off the road to change drivers or some such thing. Chris and I got out of the car; to the south the sky was as black as a sky can get, with flashes of lightning in the late July afternoon. The weather system stretched over us and maybe thirty miles to the north; directly to the west of us was a treeline, beyond which was the ocean with a hand's breadth or two of clear sky between it and the lower edge of the clouds, which was composed of gorgeous cloud lacework, brilliantly illuminated by the sun, which was about to emerge from the dark area above. In that flaming golden cloud lacework were four letters, each astonishingly legible (not razor sharp but amazingly close; far more so than anything I'd ever seen or have seen since), evenly spaced and of equal size. They spelled "G O O D". I didn't have a camera (and I wouldn't bet the farm that it would have registered in a camera!) but all three of us saw it. I strongly suspect that further along after I leave this world someone will tug at my sleeve, as it were, and say "Remember those clouds south of Fortuna?"

───────────

When I was living in Los Angeles in the mid-sixties the house I shared with a friend in Laurel Canyon was burglarized; I must have narrowly missed surprising the thief, since my best guitar (an extremely rare, Stradivarius- grade Martin from 1913) was half-way out of the house when I walked in. Seven instruments were taken, at least two of which were virtually irreplaceable, and all of which were like living beings to me not only because of their intrinsic value and the fact that they had taken me years to collect, but also because I made my living with them.

Those facts combined with the pain of violation which always comes with the failed human trust of robbery, had me hurting badly and I sat down to give it a good going over. That internal dialogue resulted in my realizing that while the thief had stolen my possessions, he had also stolen my peace of mind, and while there appeared to be little I could do about the former, I might be able to deal with the latter by thinking the thing through. I realized that when all was said and done, the best attitude I could possibly take toward the miserable son-of-a-bitch who had ripped me off would be to bless him; to pray that all that was sweet and good would somehow come to him without delay. I didn't (and don't) believe in punishment; the best I could hope for was for this person to stop doing this kind of thing, and the way for that to happen was for him to become wise and aware, so that is what I wished for him. And by the way, Lord, if I could somehow get those instruments back...

Not bloody likely in L.A. However, two weeks went by, during which time the police got nowhere. Then I got a call from the only music store I had informed of the robbery; a splendid little hole-in-the-wall in Santa Monica. They informed me that someone was coming in to offer to sell them what sounded like one of my best guitars. So it turned out to be; the kid who brought it in got it from his uncle, who got it from a pawn shop, where I recovered four more of the lost instruments; all but one of the best ones. I eventually sold that guitar for twenty-five thousand dollars.

So when we moved up here to the Northern California hills and rented a house in town to store our things while we built our house, and when we were then robbed of almost all these things I lost five more instruments, one of which (an original five-string Gibson flat-top top-tuning Mastertone banjo) was virtually irreplaceable. "I know how to deal with this!" says I, and prayed for the thief exactly as I had done in L.A. Once again I told only one music store about the robbery; a small place in Berkeley. The father of the store owner was going through music stores in Sacramento one day, and in a back street pawn shop he found a wonderful banjo, which his son realized must be mine. And so it was; I got it back.

I was once called back from across town one Friday night in Cambridge to deal with an acquaintance who had completely lost his mind. He was the kind of genius who occasionally flies too close to the edge of madness, and such was the case then. Nobody in the room (our living room) could handle him, and for some reason my roommate thought I could do it. Few prospects could have appealed to me less; when I got the call I was in an extremely relaxed state of mind (fine herb) and just beginning an evening of fascinating discussion with an old friend. Nonetheless, I agreed to try.

The further across town I got the less I liked it and by the time I had climbed up the three flights of stairs to the apartment I was completely out of gas; I felt utterly without resources to deal with a madman. I really, really didn't want to do it. So I made a deal with God. I agreed to hurl my body through the door, but the rest was up to God. I did so, and suddenly found myself completely in command of what was indeed a very difficult situation. I found exactly the right words somehow coming out of my mouth; the guy came along with me as gently as a lamb. I took him home, where he straightened out and I got an invaluable, if strenuous lesson in the power and willingness of the Cosmos to step in at the end of our individual rope when allowed to do so.

My uncle B (Bowditch) was a piano tuner with perfect pitch, who was fascinated with the beautiful old steam train whistles, to the point that he travelled around jotting down the exact notes of the whistles and then made replicas of them out of copper tubing. At the time he died he was making a copy for me of one of the Boston and Maine engines that used to fire my dreams as a child on many a country night as its whistle came over the back fields and pastures from the next town. His whistles were dispersed and to my sorrow, I was only able to lay hands on one of them.

Then one day, perhaps fifteen years after his demise, my wife and I were traveling in New Hampshire (we had lived in California for more than twenty years), where we stopped at a huge antique barn. It had an enormous number of booths in it and I decided to begin my tour over in one corner of the building. There, in the first booth I looked at, were two of Uncle B's whistles! I could see him smiling; of course I bought them.

But there's more to it; Uncle B was a very spiritually aware person. He was the only family member who didn't think I was weird when I started getting weird (the family tolerated our strange ideas because they loved us). So we had a special relationship; when I got to carrying on about the great invisible power that made the atoms twinkle, he said "Oh Peter! I'm so glad you're interested in these things! Always remember, it is a power that always wanted to work with you." A bit of excellent advice that has never left me.

I once (ah, the Sixties!) spent three days with some people in Oregon who purported to be communicating with the occupants of flying saucers, and I must say that they were either actually doing so or putting on one hell of an act. Be that as it may, their medium was a portly farm boy who was also supposed to be gifted in psychometry, doing readings on significant objects and tying them in with an individual's life.

I handed him (while he was in trance) a ring that a Hopi jeweler had made for me earlier in the sixties, with a silver cross inside a circle on a turquoise background. The ring and its symbols had great significance to me, and I rather expected him to come out with a spate of spiritual declarations, but instead he immediately said "I see a purple heart." Upon being asked in what connection, he said "I see you being shot" and when asked "In the past or in the future?" he certainly had my attention but he faltered and seemed unable to go further.

At that point a completely different voice and personality came through him (not the only time I've seen this happen). A relaxed, friendly, and intelligent voice said "Perhaps I can help. Peter was a buddy of mine during World War One. He was flying as an observer in one of those early airplanes when they were shot down by the Germans. They managed to land the plane in no-man's-land but the Germans shot them both when they emerged from the plane. I have since left the body by natural means and I have been coming to Peter (mostly in his sleep, although I did meet him in the body near here about a year ago) to impart things to him that I thought would interest him."

There were two things about his statement that seemed particularly interesting to me. One was the fact that I had always loved flying; my father had an excellent collection of early books on flight, and I devoured them. I had read several books about the aerial war in WWI; I knew every Spad, Focker, and Sopwith airplane in that war. In addition, I had indeed been in the area a year earlier; the only other time I had ever been there. While I couldn't remember meeting anyone who I could associate with the speaker, it seemed significant.

Then with this as background, I was visiting my sister in Vermont one day and as I was perusing her bookshelf I discovered a fat two-volume set from my father's collection (lent or given, depending on who you asked, by my pilot older brother to her husband, who restored Waco biplanes) listing every known aviator from New England who fought in WWI, with thumbnail biographies and photographs. Now, it wouldn't have been scientific of me not to have taken at least a brief look to see if I could find myself (since I was supposed to have been one of them), so I did. After having looked at fifty or more bios and photos I realized that there were about one hundred and fifty in each volume; no way I had time to check them all, so I decided to just look in the eyes of each aviator to see if anything clicked. Something did click when I looked at the very next photo (really!), and when I read the bio I found that he was a Bowditch (part of our family line). Hard to say what that meant, but it certainly felt meaningful.

So I'm pretty weird by any normal standard; no doubt about it. But for good reasons, it seems to me and I hope to you too.

In addition to such experiences I have had three particularly wonderful teachers regarding the Four Great Questions about what Life actually is: Jesus Christ, whose teachings, once you dig through all the various interpretations we've put on them, are unsurpassed; they ring with Truth like the Liberty Bell. Then, my personal guru; Isabel Hickey, a person of great spiritual awareness with whom a few of us fortunate hippies were able to study in the mid-Sixties. And my favorite author, Thomas Troward, whose insights and extraordinary ability to express them have five of his books by my bedside where I never stop reading them. Each time I come back to one of them I get more out of it because I bring more to it. Which realization finally allowed me to make sense of "To them that hath shall be given; to them that hath not, even that that they hath shall be taken away" (which had always seemed terribly unfair); to the extent that we keep searching out the real Truth of things, the more it reveals itself to us but the less we seek it out, the more it recedes from us.

THE TRUTH, THE WHOLE TRUTH,
AND NOTHING BUT THE TRUTH

WE CAN BE attracted to spiritual ideas but still fail to benefit significantly from them.

Ideas are photographs; if we content ourselves with admiring the pictures without going to the places that they represent, we miss out on what is by far the main point of the ideas; the living awareness of the truth that they represent. Admiring the beauty of a photograph of, say, Yosemite or the Grand Canyon is all very well but actually being there is a vastly larger experience.

Knowledge is power; spiritual realities are the real operative systems in Creation and as with any system, truly understanding them equals knowing how to use them. Our religions are the repositories of our spiritual ideas; statements of spiritual truth. We realize that these ideas are blue chips but we don't know how to cash them. We run our fingers through them like Midas with his gold because we recognize that they are valuable; very valuable, but we fail to see the nature of their deeper value. So we love the idea of eternal life but it is not real to us; we fear death. We espouse the goodness of love but we continue to hate. We give lip service to cooperation and concern for our fellow creatures but we continue to wreak the horrors of war upon each other. We talk about faith but we're filled with fear. We say we worship God ("In God We Trust") but we follow Mammon. So here we are, in truly desperate condition. We have everything we need to make life what it should be but since we haven't understood the nature of the task that is and has always been the purpose of our life ("The purpose of life is to understand the purpose of life." Sorry; I couldn't resist!), those tools are of virtually no use to us. We drift blindly ahead on the course that has led us to the brink of this cosmic Cliff, helplessly under the influence of our ignorance.

All the tools we need are at hand. Our toolbox is full, and it is magnificent. All that is needed is understanding to escape from the curse of ignorance. This dissertation has been an attempt to provide some of that understanding.

CODA

LET US FINALLY, at last, cast off this awful darkness! For so very long we have suffered through the curse that we brought upon ourselves with The Fall! How can words even hint at the pain we have suffered; pain of every imaginable kind and of such dreadful intensity? Can there be any question that after stripping away the crust of negative habits of thought that we have allowed ourselves to accumulate and to consider as not just "normal" but necessary, we will find that each of us really, deep down in what we call our hearts, wants nothing but the best for every one of us and for all the rest of God's Creation? With all the knowledge we have attained; with all the capabilities we have demonstrated, can we not yet realize what wonderful creatures we are? Can we not yet see that all the problems that dim that realization; all the not wonderful things we do, are no more nor less than symptoms of the One Great Problem that we brought into our lives with The Fall? That if and when we see this, we can attack the problem in its lair and free ourselves at last from this dreadful curse?

It is time! We have suffered long enough. We have brought things to the point where we face the necessity of choosing consciously and deliberately between Right and Wrong, understanding what we're doing and why. We must realize that Wrongness is the cause of every problem we have ever had and that Right is the solution to all of them, forever. Forever! Nothing less than that now beckons to us. All we have to do is make that choice; that is what we should always have done but we just couldn't see it. Now that choice is being put before us in the starkest terms; if we can see that and at last make that decision the Heavens will open to us; we will see that we have and will always have everything we need to do what we should be doing (because we will finally be doing what we should have been doing all along), and the very vistas of which we have scarcely dared to dream in our minds but which have always lived deep in our hearts will open to us in new worlds of wonder and glory.

That is The Choice before us.